T0133362

Test Signal Generation for Service Diagnosis Based on Local Structure Graphs

Vom Promotionsausschuss der
Fakultät für Elektrotechnik und Informationstechnik
der Ruhr-Universität Bochum
zur Erlangung des akademischen Grades
Doktor-Ingenieur
genehmigte Dissertation von

Michael Ungermann
aus Hanau

Gutachter:
Prof. Dr.-Ing. Jan Lunze
Prof. Dr. Louise Travé-Massuyès

Mündliche Prüfung:
07. Juli 2014

Bibliographic information published by the Deutsche Nationalbibliothek

The Deutsche Nationalbibliothek lists this publication in the Deutsche
Nationalbibliografie; detailed bibliographic data are available
on the Internet at http://dnb.d-nb.de .

ISBN 978-3-8325-3954-2

Logos Verlag Berlin GmbH
Comeniushof, Gubener Str. 47,
10243 Berlin
Tel.: +49 (0)30 42 85 10 90
Fax: +49 (0)30 42 85 10 92
INTERNET: http://www.logos-verlag.de

To my parents.

Preface

This thesis is the result of my research as a student at the Institute of Automation and Computer Control at the Ruhr-Universität Bochum and the department CR/AEH of Robert Bosch GmbH from August 2008 to February 2014.

The support of several people made this dissertation possible. Out of these I'd like to thank Prof. Dr. Jan Lunze first. The systematic approach to solving problems and writing structured academic texts he taught me are only some of the valuable assets I take away from this time. In particular, I would like to thank him for his excellent supervision, his patience and his kind questions concerning my progress in writing up this thesis.

I would also like to thank Prof. Dr. Louise Travé-Massuyès for being the second reviewer of this work and her valuable suggestions concerning this text.

Special thanks go to Dr. Dieter Schwarzmann who gave me orientation in both, the company and in the academic world. His advice when it came to publishing results and applying for patents was of inestimable value.

I also thank my fellow students at the Institute of Automation and Computer Control and at Robert Bosch GmbH. The lively discussions we had about various subjects in the field of control theory contributed a lot to this thesis. In particular, the open-mindedness of Philipp Kotman and Dr. Michael Hilsch was a great help.

Further thanks go to my students Markus Koslowski and Fanny Djoutsop for their valuable support in forging new algorithms into a MATLAB toolbox and in simulating new active diagnostic approaches.

On a personal level, I would like to thank my family and my friends for their patience during the time I worked on this thesis.

Thank you.

Ludwigsburg, August 2014   Michael Ungermann

Contents

Abstract

The goal of service diagnosis is the identification of the faulty component in a defective system. The thesis at hand describes a novel approach to the design of automatic tests which can be used for this purpose. The tests provided by the new method use the input and output signals of the system in order to reject inapplicable fault-hypotheses, thus narrowing a set of fault candidates. The consecutive execution of such tests in the sequential process of service diagnosis eventually leads to the fault which is present in the system. In that way, the faulty component can be identified.

Automatic tests lead to active diagnosis. In contrast to process diagnosis, they use dedicated system excitation to reject wrong fault-hypotheses. The main idea behind automatic tests consists of steering the system into operating regions which are beneficial to the distinction of faults. Then the consistency of the input and output signals of the plant with its fault-free behavior is checked in a diagnostic unit. Methods for the determination of such operating regions, the generation of the steering commands, as well as methods to determine the algorithms which implement the consistency tests are developed in this thesis.

These methods use an extension of the structure graph of a model of the faulty system. In the structure graph, couplings among the variables in the model are represented by edges between vertices representing variables and constraints respectively. The extension suggested in this thesis concerns the case where such couplings vanish in specific operating regions. If the system is restricted to such an operating region, the corresponding structure graph differs from the one present in different operating regions.

This new *local structure graph* allows to determine the components of the diagnostic unit. These components are dynamical systems which compute two kinds of signals from the plant's input and output signals: the *local residuals* which realize consistency tests in specific operating regions, and the *validuals* which indicate the presence of such an operating region. Comparing these signals to thresholds allows a decision logic to reject wrong fault-hypotheses. This logic is likewise determined by the help of structural analysis.

The method to determine input generators providing signals that steer the system into a specific operating region is based on the validuals. Since they indicate the presence of an operating region, the test signals are obtained by steering the validuals in the desired way.

An automatic test can be constructed from an input generator and a diagnostic unit, all of which can be obtained by the methods developed in this thesis.

These methods are illustrated by their application to a typical automotive system, a throttle valve. It is shown that the resulting test allows to distinguish faults which cannot be distinguished using other structural approaches.

Keywords: Structural analysis, service diagnosis, active diagnosis, test generation, automotive application.

Deutsche Kurzfassung (German Extended Abstract)

Motivation und Zielsetzung

Die Werkstattdiagnose hat das Ziel, die fehlerbehaftete Komponente in einem defekten technischen System zu ermitteln. Diese, für die Reparatur des Systems notwendig Aufgabe, wird durch die zunehmende Komplexität und Verkopplung moderner technischer Systeme erschwert. Eine in Kap. 2 beschriebene Vorgehensweise, die defekte Komponente zu bestimmen besteht darin, ausgehend von einer Menge von *Fehlerhypothesen*, eine Reihe von *Tests* auszuführen. Die Tests verwerfen jeweils unzutreffende Fehlerhypothesen und verfeinern auf diese Weise das Diagnoseergebnis schrittweise. Im günstigen Fall bleibt am Ende dieses sequenziellen Prozesses lediglich eine Fehlerhypothese übrig, die nicht falsifiziert werden konnte. Diese entspricht dem tatsächlich vorliegenden Fehler. Eine Zuordnung von Fehlern zu Komponenten erlaubt es, mit diesem Ergebnis die defekte Komponente zu bestimmen.

Automatische Tests verfolgen den Ansatz, das zu untersuchende System in geeigneter Weise anzuregen und unzutreffende Fehlerhypothesen lediglich auf Basis des Eingangs-/ Ausgangsverhaltens des Systems zu verwerfen. Das Blockschaltbild eines solchen Tests ist in Abb. 1 angegeben.

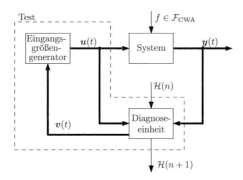

Abbildung 1: Blockschaltbild eines automatischen Tests

Ein *Eingangsgrößengenerator* regt das System mit einem spezifischen Eingangsgrößenverlauf an. Dieser Verlauf und der Ausgangsgrößenverlauf des Systems werden von einer *Diagnoseeinheit* zum Verwerfen unzutreffender Fehlerhypothesen verwendet. Typischerweise haben Fehler in bestimmten Betriebszuständen einen deutlichen, in anderen nur einen schwachen oder gar keinen Einfluss auf das System. Dieser Sachverhalt kann durch automatische Tests für Diagnosezwecke genutzt werden. Die vorliegende Arbeit behandelt daher die Fragestellungen:

- Wann existiert ein automatischer Test?

- Wie muss das System zur Diagnose angeregt werden?

- Wie kann die Diagnoseeinheit eines automatischen Tests bestimmt werden?

- Unter welchen Bedingungen können welche Hypothesen falsifiziert und so das Diagnoseergebnis verfeinert werden?

Lösungsweg

Der hier gewählte Ansatz fällt in die Klasse der Verfahren der *aktiven Diagnose*, bei denen ein System gezielt zu Diagnosezwecken angeregt wird. Im Unterschied zu den aus der Literatur bekannten Verfahren kommt weder ein stochastischer Ansatz [71, 72, 130], noch ein Ansatz, der ein lineares System voraussetzt [84, 86, 91] zum Einsatz. Vielmehr wird das z.B. aus [22, 76] für passive Diagnose bekannte Konzept des Strukturgraphen verwendet und für die aktive Diagnose erweitert, um die oben genannten Fragestellungen im Kontext deterministischer, nichtlinearer dynamischer Systeme mit bekannten Anfangsbedingungen zu beantworten.

Konsistenztest mit Eingangs-/Ausgangsverhalten. Der Lösungsweg basiert auf dem Konzept des *Konsistenztests*. Bei diesem Test wird überprüft, ob ein *Eingangs-/Ausgangsgrößenverlaufspaar*, das am zu untersuchenden System beobachtet wurde, im fehlerfreien Verhalten des Systems enthalten ist. Ist dies nicht der Fall, wird auf das Vorliegen eines Fehlers geschlossen. Bei dem in Abschn. 3.2 eingeführten Begriff des *Verhaltens* eines deterministischen Systems handelt es sich um die Menge aller möglichen Eingangs-/Ausgangsgrößenverlaufspaare des Systems. Es kann sich je nach Fehlerzustand des Systems verändern. In Abb. 2 sind das Verhalten eines Systems ohne Fehler, in verschiedenen Fehlerfällen, sowie verschiedene mit den Verhalten konsistente und inkonsistente E/A-Paare dargestellt.

Bestimmung globaler Residuengeneratoren mit dem globalen Strukturgraphen. In Abschn. 4.2 wird beschrieben, wie sich mit Hilfe *globaler Residuengeneratoren* Konsistenztests realisieren lassen. Globale Residuengeneratoren sind dynamische Systeme, die aus den Eingangs- und Ausgangssignalen eines potenziell fehlerhaften Systems *globale Residuen* genannte Signale $r(t)$ ermitteln. Wenn ein globales Residuum ungleich null ist, ist das E/A-Paar inkonsistent mit dem fehlerfreien Verhalten und es liegt ein

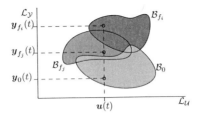

Abbildung 2: Eingangs-/Ausgangsverhalten eines dynamischen Systems

Fehler im System vor. Ist das System fehlerfrei, sind die globalen Residuen für beliebige Eingangssignale null:

$$r(t) \neq 0 \quad \Rightarrow \quad \exists\; f_i \neq 0$$
$$f_i = 0 \;\forall\; i \quad \Rightarrow \quad r(t) = 0 \;\forall\; t. \tag{1}$$

Der eigentliche Konsistenztest wird durch den Vergleich des globalen Residuums mit einem Schwellenwert vorgenommen. Mehrere globale Residuengeneratoren und eine Bool'sche Entscheidungslogik erlauben das Unterscheiden von Fehlern und damit das Ausschließen unzutreffender Fehlerhypothesen. Grundsätzliche Eigenschaften globaler Residuen werden in Abschn. 4.2 beschrieben. Die Eigenschaften fehlerhafter Systeme, die prinzipiell die Detektion eines Fehlers bzw. die Unterscheidung von zwei Fehlern erlauben, werden in Abschn. 4.3-4.4 untersucht.

Ein Verfahren, um globale Residuengeneratoren und eine Entscheidungslogik zu bestimmen, ist die Analyse des *globalen Strukturgraphen G*, der die Kopplungen in einem technischen System in Form eines bipartiten Graphen darstellt. Er wird in Abschn. 4.5 als

$$G = (\mathcal{C} \cup \mathcal{Z}, \mathcal{E}) \quad \text{mit} \quad \mathcal{Z} = \mathcal{X} \cup \mathcal{K} \cup \mathcal{F} \tag{2}$$

definiert, wobei die Knoten des Graphen die Modellgleichungen $\mathcal{C}$ und die in ihnen vorkommenden Variablen $\mathcal{Z} = \mathcal{X} \cup \mathcal{K} \cup \mathcal{F}$ sind. Letztere entsprechen den unbekannten (Zwischen-) Variablen $\mathcal{X}$, den bekannten Variablen $\mathcal{K}$, d.h. Ein- und Ausgängen des Systems, und den Fehlervariablen $\mathcal{F}$, die den Fehlerzustand des Systems beschreiben. Im globalen Strukturgraphen gibt es Kanten $\mathcal{E}$ zwischen einem Variablen- und einem Gleichungsknoten genau dann, wenn die Variable in der Gleichung vorkommt. Ein Beispiel eines globalen Strukturgraphen ist in Abb. 3 gezeigt.

Mit der in Abschn. 4.5-4.7 beschriebenen Analyse des globalen Strukturgraphen können Gleichungsmengen bestimmt werden, aus denen durch Eliminieren der unbekannten Variablen globale Residuengeneratoren gewonnen werden können. Diese Gleichungsmengen heißen global minimal strukturell überbestimmt (GMSO). Der globale Strukturgraph erlaubt es darüberhinaus zu ermitteln, welche Fehler dazu führen können, dass ein globales Residuum ungleich null ist. In Abschn. 4.8 münden diese Aussagen in den Eigenschaften der globalen strukturellen Detektierbarkeit eines Fehlers und der globalen strukturellen Diskriminierbarkeit von zwei Fehlern. Sie geben an, ob für den Fall, dass

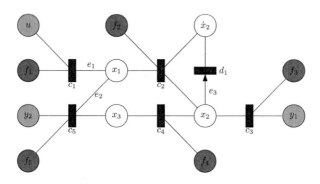

Abbildung 3: Globaler Strukturgraph

mit der beschriebenen Vorgehensweise ermittelte globale Residuen ungleich null sind, eine Bool'sche Entscheidungslogik es erlaubt einen Fehler zu erkennen, bzw. zwei Fehler voneinander zu unterscheiden.

Der Zusammenhang zwischen der globalen strukturellen Detektierbarkeit und Diskriminierbarkeit der Fehler und der Eigenschaft einen Fehler am Eingangs-/Ausgangsverhalten tatsächlich erkennen bzw. zwei Fehler anhand des Verhaltens unterscheiden zu können, wird ebenfalls in Abschn. 4.8 und Abschn. 4.9 untersucht.

Globale Residuengeneratoren bestimmen globale Residuen mit den in Gl. 1 beschriebenen Eigenschaften für beliebige Eingangs- und Ausgangsgrößenverläufe. Sie sind daher in besonderem Maß zur Überwachung technischer Systeme, bei der es vornehmlich um das Erkennen von Fehlern während des Betriebs und erst im zweiten Schritt um deren Unterscheidung geht, geeignet.

Bestimmung lokaler Residuengeneratoren mit dem lokalen Strukturgraphen. Die Anzahl der GMSOs, die mit der Analyse des globalen Strukturgraphen gefunden werden können, ist begrenzt und typischerweise kleiner als die Anzahl der möglichen Fehler. Aus diesem Grund kann mit globalen Residuen ein vorliegender Fehler zwar erkannt, aber häufig nicht eindeutig bestimmt werden. Das ursprüngliche Ziel, die defekte Komponente zu bestimmen, kann in diesem Fall nicht erreicht werden.

Da sich das System während der Werkstattdiagnose nicht in Betrieb befindet, kann es zu Diagnosezwecken gezielt in spezifische Betriebsbereiche gesteuert werden, ohne dass das ursprüngliche Ziel des Systems wie z.b. die Sollwertfolge erfüllt werden muss. Steuert ein Eingangsgrößenverlauf das System in einen solchen Betriebsbereich, so kann ein Konsistenztest in diesem Betriebsbereich realisiert werden, ohne dass eine zentrale Eigenschaft des globalen Residuums gilt. Anstatt $r(t) = 0 \ \forall \ t$ für beliebige Eingangs- und Ausgangssignale im fehlerfreien Fall reicht es aus, wenn diese Eigenschaft für die Eingangsgrößenverläufe, die das System in den spezifischen Betriebsbereich steuern, erfüllt ist. Die Anforderungen an einen Konsistenztest in einem spezifischen Betriebsbereich sind also weniger scharf als die an das globale Residuum.

Ist ein Betriebsbereich durch eine Menge von Zwangsbedingungen $\mathcal{C}_w$ für die bekannten und unbekannten Variablen definiert, so lässt sich ein Konsistenztest auch mit Hilfe des in Abschn. 5.6 eingeführten *lokalen Residuums* $r_w(t)$ realisieren. Dabei handelt es sich um ein Signal, das null ist, wenn das System fehlerfrei ist und sich darüberhinaus in einem spezifischen Betriebsbereich befindet. Ist das lokale Residuum ungleich null und befindet sich das System in dem spezifischen Betriebsbereich, so liegt ein Fehler vor:

$$r_w(t) \neq 0 \text{ und System in } \mathcal{C}_w \Rightarrow \exists f_i \neq 0$$
$$f_i = 0 \ \forall \ i \text{ und System in } \mathcal{C}_w \Rightarrow r_w(t) = 0 \ \forall \ t. \tag{3}$$

Dynamische Systeme, die aus den Eingangs- und Ausgangssignalen des zu untersuchenden Systems lokale Residuen berechnen, werden *lokale Residuengeneratoren* genannt. Zur Bestimmung von zu Diagnosezwecken besonders geeigneten Betriebsbereichen und lokalen Residuengeneratoren wird von der folgenden, in Abschn. 5.3 detailliert beschriebenen Beobachtung ausgegangen:

Im globalen Strukturgraphen kommen Kanten zwischen einer Variable z_j und einer Gleichung c_i vor, wenn die Variable in die Gleichung eingeht. Dies ist auch der Fall, wenn in bestimmten Betriebsbereichen eine Veränderung des Wertes der Variable z_j den Zusammenhang, den die Gleichung c_i zwischen den anderen Variablen herstellt, nicht verändert. Unter der Annahme, dass sich das System in einem solchen durch die Zwangsbedingung $c_{\mathrm{Elim},k}$ beschriebenen Betriebsbereich befindet, kann die Kante zwischen z_j und der Gleichung c_i aus dem Strukturgraphen entfallen. Diese Kante heißt lokal unwirksam unter der Bedingung $c_{\mathrm{Elim},k}$. Die Untersuchung aller Kanten im globalen Strukturgraphen erlaubt es, Betriebsbereiche, d.h. Zwangsbedingungen für die Variablen zu bestimmen, in denen Kanten lokal unwirksam sind.

Diese Betrachtung zeigt, dass der Strukturgaph vom Betriebsbereich des Systems abhängig ist und erlaubt die Definition des *lokalen Strukturgraphen* in Abschn. 5.4. Dabei handelt es sich um einen Graphen, der die Kopplungen des Systems beschreibt, wenn dieses in den entsprechenden Betriebsbereich gebracht wurde. Der lokale Strukturgraph wird aus dem globalen Strukturgraphen gebildet, indem lokal unwirksame Kanten $\mathcal{E}_w$ weggelassen werden und die Zwangsbedingungen $\mathcal{C}_w$, die den entsprechenden Betriebsbereich beschreiben in den Graphen eingefügt werden:

$$G|_{\mathcal{C}_w} = \{(\mathcal{C} \cup \mathcal{C}_w) \cup \mathcal{Z}, \mathcal{E} \setminus \mathcal{E}_w\}. \tag{4}$$

Ein Beispiel eines lokalen Strukturgraphen ist in Abb. 4 dargestellt. Dieser Graph unterscheidet sich von dem globalen Strukturgraphen in Abb. 3 dadurch, dass die Kante e_2 nicht in ihm vorkommt und die Zwangsbedingung $c_{\mathrm{Elim},2}$, die den Betriebsbereich beschreibt, hinzugefügt wurde.

Unter der Annahme, dass sich das System tatsächlich in diesem Betriebsbereich befindet, beschreibt der lokale Strukturgraph die tatsächlich wirksamen Kopplungen im System. Er erlaubt es deshalb, Konsistenztests der Eingangs- und Ausgangssignale des Systems mit dessen nominalen Verhalten in diesem Betriebsbereich in Form von lokalen Residuengeneratoren zu bestimmen.

Mit der Analyse des lokalen Strukturgraphen mit den Verfahren, die aus der Analyse des globalen Strukturgraphen bekannt sind, können Gleichungsmengen gefunden werden, die minimal strukturell überbestimmt auf dem lokalen Strukturgraphen sind. Werden aus einer solchen lokal minimal strukturell überbestimmten Gleichungsmenge (LMSO)

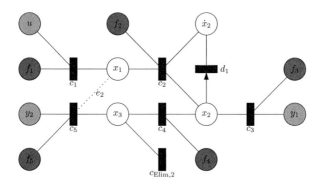

Abbildung 4: Lokaler Strukturgraph

die unbekannten Variablen eliminiert, so erhält man einen lokalen Residuengenerator. Im Allgemeinen ergibt die lokale strukturelle Analyse mehr lokale Residuengeneratoren als die globale strukturelle Analyse zu globalen Residuengeneratoren führt. Dies legt die Vermutung nahe, dass mit lokalen Residuen Fehler besser voneinander unterschieden werden können als mit globalen Residuen, und es deshalb erlauben, die ursprüngliche Aufgabe der Bestimmung der defekten Komponente zu lösen.

Validuum. Der in Kap. 6 formulierte Vorschlag ist daher, ein zu diagnostizierendes System gezielt so zu steuern, dass ein spezifischer lokaler Strukturgraph gilt. Dann ermöglichen die entsprechenden lokalen Residuen einen Konsistenztest und können in einem automatischen Test, der unzutreffende Fehlerhypothesen verwirft, verwendet werden.

Allerdings muss dazu ein potenziell fehlerhaftes System erfolgreich in einen spezifischen Betriebsbereich gesteuert werden, was im Allgemeinen schwierig ist. Ein einfacherer, in dieser Arbeit verfolgter Ansatz besteht darin, stattdessen das Vorliegen eines solchen Betriebsbereichs zu detektieren. Dazu wird in Abschn. 6.4 das *Validuum* $v_k(t)$ eingeführt. Dabei handelt es sich um ein Signal, das aus den Eingangs- und Ausgangssignalen des Systems bestimmt wird. Es hat die Eigenschaft, dass wenn das Validuum $v_k(t)$ null ist, die Zwangsbedingung $c_{\text{Elim},k}$ erfüllt ist. Dann ist die Kante e_k des globalen Strukturgraphen lokal unwirksam. Ein Betriebsbereich, in dem ein lokaler Strukturgraph gilt, in dem mehrere Kanten des globalen Strukturgaphen lokal unwirksam sind, kann erkannt werden wenn alle entsprechenden Validuen null sind.

Die Validuen können mit Hilfe von GMSOs eines spezifischen globalen Strukturgraphen bestimmt werden. Dieser globale Strukturgraph beinhaltet neben den Gleichungen, die das System beschreiben, auch die Zwangsbedingung $c_{\text{Elim},k}$, die den Betriebsbereich beschreibt, in der die Kante e_k lokal unwirksam ist.

Diagnoseeinheit. Das Blockschaltbild der in Abschn. 6.3 eingeführten Diagnoseeinheit eines automatischen Tests ist in Abb. 5 dargestellt. Sie besteht aus der Detektion eines Betriebsbereichs mit Hilfe von Validuen, einem Konsistenztests auf Basis eines lokalen Residuengenerators und einer Bool'schen Logik, die entscheidet, welche Fehlerhypothesen

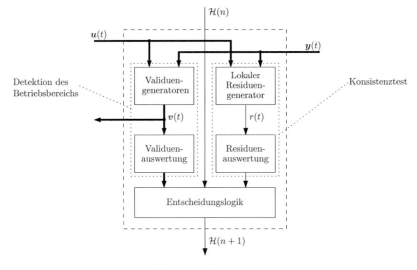

Abbildung 5: Diagnoseeinheit

verworfen werden. Die Lösung des schwierigen Problems, ein potenziell fehlerhaftes System zu Diagnosezwecken gezielt zu steuern, besteht darin, die Detektion des Betriebsbereichs für das fehlerfreie System auszulegen und in der Entscheidungslogik diejenigen Fehler zu berücksichtigen, die den Schluss auf den vorliegenden Betriebsbereich verfälschen können.

Dies führt auf das Ergebnis in Satz 6.1, dass wenn alle Validuen, die einen bestimmten Betriebsbereich anzeigen null sind und ein lokales Residuum in diesem Betriebsbereich ungleich null ist, nicht nur solche Fehler vorliegen können, die die LMSO betreffen mit der das lokale Residuum bestimmt wurde. Vielmehr kann auch einer derjenigen Fehler vorliegen, die die GMSOs betreffen, aus denen die Validuen gebildet wurden. Beide Fehlergruppen können mit Hilfe der lokalen und der globalen strukturellen Analyse bestimmt werden. Dies erlaubt es zu bestimmen, welche Fehlerhypothesen verworfen werden können.

Eingangsgrößengenerator. Da von den Validuen ein Betriebsbereich angezeigt wird, wenn diese null sind, kann der Eingangsgrößengenerator eines Tests bestimmt werden, indem die Validuen null gesetzt und nach den Eingängen aufgelöst werden. Dieser, in Abschn. 6.6 vorgestellte Ansatz, hat zur Folge, dass sich die Eingangsgrößengeneratoren bzgl. ihrer regelungstechnischen Struktur unterscheiden. Sie hängt davon ab, ob nur Eingangs-, nur Ausgangs-, oder Eingangs- und Ausgangsvariablen in den GMSOs, die zur Berechnung der Validuen verwendet werden, vorkommen. Diese Eigenschaften können ebenfalls mit Hilfe der strukturellen Analyse gezeigt werden.

Ergebnisse

Die einzelnen Erkenntnisse und Verfahren der vorliegenden Arbeit wurden anhand mehrerer Beispiel erprobt und die Wirksamkeit der resultierenden Tests simulativ untersucht. Die Verfahren und Ergebnisse wurden in mehreren Publikationen veröffentlicht [2, 3, 4, 5, 6]. Einzelne spezifische Tests wurden zum Patent angemeldet [1, 7, 8, 9, 10]. In der vorliegenden Arbeit werden die Verfahren genutzt, um am Beispiel einer Drosselklappe einen automatischen Test zu erzeugen. Die wesentlichen Beiträge dieser Arbeit sind:

- Das Konzept des automatischen Tests, der mit Hilfe gezielter Anregung und eines Konsistenztests der Eingangs-/ und Ausgangssignale des Systems mit dessen nominalen Verhalten unzutreffende Fehlerhypothesen verwirft und sich so in den sequenziellen Prozess Werkstattdiagnose einbinden lässt.

- Die Erkenntnis, dass sich der Strukturgraph eines Systems mit dem Betriebszustand verändern kann und der daraus abgeleitete lokale Strukturgraph. Dieser beschreibt die Kopplungen im System, wenn sich das System in einem spezifischen Betriebszustand befindet.

- Das Validuum, ein Signal, das es erlaubt auf das Vorliegen eines Betriebsbereichs zu schließen, in dem eine Kante lokal unwirksam ist und ein Verfahren, mit dem sich Validuengeneratoren bestimmen lassen.

- Das lokale Residuum, das unter der Prämisse, dass sich das System in einem spezifischen Betriebsbereich befindet einen Konsistenztest realisiert, sowie ein Verfahren, mit dem sich lokale Residuengeneratoren bestimmen lassen.

- Die Integration von Validuengeneratoren, einem lokalen Residuengenerator und einer Entscheidungslogik zu einer Diagnoseeinheit, die als Teil eines automatischen Tests unzutreffende Fehlerhypothesen verwirft.

- Ein Verfahren zur Bestimmung von Eingangsgrößengeneratoren, sowie die Bestimmung der regelungstechnischen Struktur eines Eingangsgrößengenerators mit Hilfe der strukturellen Analyse.

Nomenclature, Abbreviations and Symbols

In the following, an overview of the notation used in this thesis is given.

Symbols

Tests and Hypotheses
T Test
h_i fault-hypothesis i
$\mathcal{H}$ set of fault-hypothesis

Signals
$\boldsymbol{u}(t)$ input signal
$\boldsymbol{y}(t)$ output signal
$r(t)$ residual
$r_w(t)$ local residual in $\mathcal{C}_w$
$v_k(t)$ validual indicating the local ineffectiveness of edge e_k

Variables
$\mathcal{Z}$ set of variables
$\mathcal{X}$ set of unknown variables
$\mathcal{U}$ set of input variables
$\mathcal{Y}$ set of output variables
$\mathcal{K}$ set of known variables
$\mathcal{F}$ set of fault variables
f_i fault i

Constraints and constraint sets
c_i name of constraint i
g_i constraint c_i in residual form
$\mathcal{C}$ arbitrary constraint set, model description
$\mathcal{C}_w$ description of operating region w
$\mathcal{C}_U$ subset of the constraint sets $\mathcal{C}$ or $\mathcal{C} \cup \mathcal{C}_w$
$c_{\text{Elim},k}$ operating region in which edge the e_k is locally ineffective

Behavior

$\mathcal{B}$ behavior

$\mathcal{B}(\mathcal{C})$ behavior defined by the constraint set $\mathcal{C}$

$\mathcal{B}_0$, $\mathcal{B}(\mathcal{C}_0)$ nominal behavior

$\mathcal{B}_{f_i}$, $\mathcal{B}(\mathcal{C}_{f_i})$ behavior in the case of fault f_i

$\mathcal{B}(r(t) = 0)$ I/O-pairs which lead to the residual $r(t)$ being zero

Structure graphs

G global structure graph

$G(\mathcal{C})$ global structure graph of $\mathcal{C}$

$G|\mathcal{C}_w$ or $G(\mathcal{C})|\mathcal{C}_w$ local structure graph of $\mathcal{C}$ in the operating region $\mathcal{C}_w$

Edges

e edge

$\mathcal{E}$ set of edges

$\mathcal{E}_w$ set of locally ineffective edges

$\mathcal{M}$ matching

Operators

$|\cdot|$ cardinality of a set

$\frac{d}{dt}$ indicates that the derivatives of signals w.r.t. time of arbitrary order may occur

$\mathrm{var}^G(\mathcal{C})$ variables which have an impact on one of the constraints in the set $\mathcal{C}$ according to the structure graph G

$DM(\cdot)$ Dulmage-Mendelsohn-Decomposition

$\bar{\varphi}$ degree of structural redundancy

Abbreviations

GSO - globally structurally overconstrained

GPSO - globally proper structurally overconstrained

GMSO - globally minimal structurally overconstrained

GSJC - globally structurally justconstrained

LPSO - locally proper structurally overconstrained

LMSO - locally minimal structurally overconstrained

Figures

Diagnostic flow charts

Squares: hypothesis-sets

Circles: tests

Block diagrams

Bold lines: vector-valued signals

Fine lines: scalar-valued signals, hypothesis-sets

Structure graphs
Circle-shaped vertices: variables
  - white: unknown variables
  - light gray: known variables
  - dark gray: fault variables
Bar-shaped vertices: constraints
Continous lines: locally effective edges
Dashed lines: locally ineffective edges
Arrows: causal edges

Algorithms

FINDMSO
Determines minimal structurally overconstrained subsets of the constraint set $\mathcal{C}$ according to a structure graph, p. 66.

DETERMINECAUSALMATCHING
Determines a causal matching on a directed structure graph, p. 73.

FINDLOCALRESIDUALS
Determines local residuals with the help of a directed local structure graph of the plant in a specific operating region, p. 109.

FINDVALIDUALS
Determines validuals with the help of a directed global structure graph, p. 120.

FINDTESTS
Determines automatic tests and their sensitivity to faults, p. 132.

Chapter 1

Thesis Overview

1.1 Service Diagnosis

Service diagnosis is the process to localize the faulty component in a defective system. It is carried out prior to the repair of the system in order to determine which component has to be replaced. Service diagnosis is a sequential process in which the diagnostic result is refined stepwise by the consecutive execution of tests. A specific kind of tests are automatic tests which identify the fault by applying dedicated test signals and analyzing the resulting input and output signals of the plant, c.f. Fig. 1.1. Automatic tests are therefore a kind of active diagnosis.

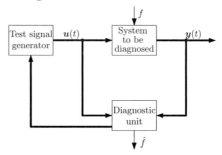

Figure 1.1: Automatic test based on active diagnosis

Automatic tests solve the cost, reliability, and duration issues of manual tests which are carried out by service technicians. However, the lack of methodical approaches for the design of automatic tests still prohibits their widespread application.

Therefore, the goal of this thesis is to provide a method of determining automatic tests for the service diagnosis of mechatronic systems. This method allows to determine

- the test signals and

- some measures which allow the localization of faults by systematic evaluation of the system's input and output signals

on the basis of an appropriate description of the system to be diagnosed.

1

1.2 Contributions of This Thesis: Systematic Test Design Methods

The contribution of this thesis is a method for the design of automatic tests which can be used in the framework of service diagnosis. The resulting tests reject wrong fault-hypotheses. This allows the refinement of the diagnostic result in the sequential process of service diagnosis. The basis of the hypothesis rejection is the dedicated excitation of the system to be diagnosed and a consistency test of the observed input and output signals with the behavior of the fault-free system.

The design method uses an extension of the well-known structure graph. The structure graph is a bipartite graph which represents the dependencies of variables and constraints. Because the structure graph is valid in any operating point, it is here called *global structure graph*. The following gives a brief overview of the main contributions of this thesis.

1. The first main contribution of this thesis is a method for the identification of operating regions in which specific cause-effect relations of a system are interrupted. It is based on the observation that some variables, although adjacent to a constraint in the global structure graph, may not have any impact on that constraint in specific operating regions. By analyzing the relation described by an edge in the global structure graph, it is possible to determine constraints describing such operating regions. If the system is restricted to an operating region represented by such additional constraints, edges in the global structure graph vanish and additional constraints appear in it, thus forming a new graph. This graph, which is called *local structure graph*, is a new tool defined in this thesis. It can be used to derive signals which realize consistency tests of the input and output signals with the behavior of the fault-free system in a specific operating region. These signals, called *local residuals*, allow the fault detection and distinction, if the system is in the corresponding operating region.

2. The second main contribution of this thesis is a method which allows to infer whether the system currently is in a specific operating region. It is based on a new kind of signal, called *validual*. These signals, which can be obtained from the system's inputs and outputs, indicate whether the system has been brought to a specific operating region. They validate that a particular edge vanishes from the global structure graph. For that reason, a set of validuals can be used to determine whether the system is currently described by a specific local structure graph. A method is given to determine constraint sets which can be used to compute validuals. This method is based on the analysis of the global structure graph. A possibly faulty system can be steered into a specific operating region by steering the validuals in a desired way. In that way, the test signals which are used for diagnosis can be determined from the validuals. Depending on the variables in the constraint sets used to determine the validuals, the resulting test signal generator can be realized by a feedforward controller, a feedback controller or is itself a control law.

3. The third main contribution of this thesis is the interconnection of validuals, a local residual, and a decision logic to a diagnostic unit. Such a diagnostic unit and an appropriate test signal generator form an automatic test. The result is an active diagnosis scheme in which dedicated test signals are applied to the system and wrong fault-hypotheses are rejected. In that way an automatic test refines the diagnostic result.

In summary, a method for the design of automatic tests which consist of a test signal generator and a diagnostic unit is developed. This method is based on the selection of an operating region, validuals, and a local residual. The selection uses the investigation of the hypothesis rejection properties of the resulting test by means of both, global and local structure graphs.

Finally, the new method is used to generate a test for detecting and distinguishing faults in a throttle valve.

1.3   Thesis' Outline

The thesis is organized as follows: In Chapter 2, the sequential process of service diagnosis is introduced and the use of automatic tests is motivated. Differences between automatic tests and existing concepts of active diagnosis are analyzed. In Chapter 3, the kind of models used by the methods in this thesis is reviewed. The analysis of the diagnosability of faults by means of global structure graphs is presented in Chapter 4. The new local structure graph and its application to the analysis of the diagnosability of faults in dynamical systems is introduced in Chapter 5. These results are used in Chapter 6 in a method for the design of automatic tests which comprise the dedicated excitation of the system to be diagnosed and the analysis of its input an output signals. The new method for the design of automatic tests which is the main contribution of this thesis, is exemplified with a throttle valve in Chapter 7. In Chapter 8, a summary of the thesis is given and fields of future research are suggested.

In the following, the first section of a chapter contains an overview of the train of thought of the chapter. The last section summarizes the results of the chapter.

Chapter 2

Introduction to Consistency-Based Service Diagnosis

2.1 Introduction to Service Diagnosis

Service diagnosis aims at identifying the faulty component in a defective system. It is applied to mechatronic systems in general and automotive systems in particular. Service diagnosis is usually carried out in a repair shop prior to the repair of the system. This is done in order to determine which component has to be replaced to restore the original functionality of the system. It solves the problem:

"Given a faulty system, identify the faulty component."

In service diagnosis, the approach to solve this problem is a sequential process, in which the preliminary diagnostic result is refined step by step. Eventually, this leads to the actually present fault or a small set of remaining fault candidates which contains the actual fault. This approach is depicted in Fig. 2.1.

Starting from a number of possible faults, the fault-hypotheses h_i in the set $\mathcal{H}(0)$, tests T_i are executed in order to rule out those fault-hypotheses which can be proven to be wrong. In that way a sequence $T(n)$ of tests results. During this test-sequence, a sequence of shrinking fault-hypothesis-sets $\mathcal{H}(n)$ is obtained. This reveals the fault which is actually present in the system. An example of such a sequence is marked by the bold arrows Fig. 2.1. Because each fault can be associated to a component of the faulty system which is referred to as *plant* in the following, determining which fault is present solves the original problem.

Today, the common practice is to apply manual tests which require repair shop personnel to intervene. This is an expensive and error-prone approach. An even more severe problem is the rising complexity of modern mechatronic systems. It generally entails a more complex system-behavior, in which it is difficult to diagnose a fault. In particular, heuristic approaches to identify operating regions in which faults are easy to diagnose and rules which link measurements of the system outputs to diagnostic results are often incomplete or fail.

5

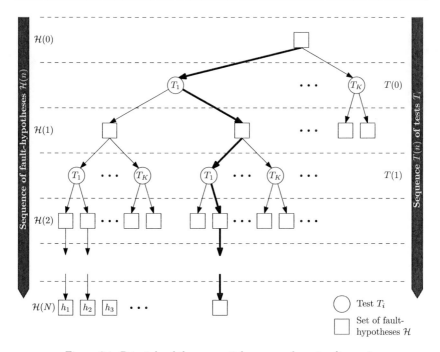

Figure 2.1: Principle of the sequential process of service diagnosis

The new approach in this thesis is to develop automatic tests with a methodical approach based on a model of the plant. Automatic tests excite the plant with specific test signals and use the input and output signals of the system to reject wrong fault-hypotheses. They are therefore based on a kind of active diagnosis.

Similar to process diagnosis, the automatic tests diagnose the plant by checking the consistency of the input and output signals observed at the plant with a model of the behavior of the fault-free plant. However, a major difference with respect to process diagnosis is that the plant is not required to meet the original functional goal during an automatic test. Instead of applying small auxiliary signals to the plant under normal operating condition, the test signals are allowed to steer the plant into arbitrary operating points. This is an important observation, which entails a main idea of this thesis: Steering into specific operating points may allow to distinguish faults. This is especially promising because faults may change the behavior of a plant in some operating points significantly, but may barely have an effect in other operating points. It therefore forms the basis of automatic tests.

In this chapter, the concepts of fault-hypotheses and their rejection by tests are formalized in Section 2.2. The formal definition of a test is used to detail the sequential process of service diagnosis in Section 2.3. Section 2.4 and Section 2.5 develop the idea of automatic tests which use dedicated test signals and model-based consistency tests. This idea is brought into the overall context of (active) diagnosis in an overview of the literature in Section 2.6. Finally, in Section 2.7 the idea of automatic tests, assumptions on the situation in the repair shop and the formal definition of a test are used for the formulation of a problem statement. This problem statement is:

"Given a model of a system, find a test signal generator and a diagnostic unit forming an automatic test which rejects wrong fault-hypotheses."

2.2 Fault-Hypothesis Rejection

In this section the concept of tests which realize the rejection of wrong fault-hypotheses and which can therefore be used in service diagnosis, is introduced.

Instead of trying to determine the fault directly, a usual approach in service diagnosis is to rule out all faults that are not present, thus narrowing down the fault to the actually present one. Ruling out a fault is the same as proving the *hypothesis* that this fault is present, to be wrong. This motivates the following definition:

Definition 2.1 (Fault-hypothesis). *The hypothesis that fault f_i is present, is denoted by h_i. The set of fault-hypotheses corresponding to the set of faults $\mathcal{F}$ is denoted by $\mathcal{H}$.*

Assuming that the plant is known to be faulty, the service diagnosis starts with the hypothesis-set

$$\mathcal{H}(0) \tag{2.1}$$

which is also called *initial guess*. This hypothesis-set contains the hypotheses corresponding to the faults which may be present in the system. Because the plant is known to be faulty, at least one of the hypotheses in $\mathcal{H}(0)$ must be true. By rejecting wrong fault-hypotheses, the process service diagnosis results in a small set of fault-hypotheses

$$\mathcal{H}(N), \quad |\mathcal{H}(N)| << |\mathcal{H}(0)| \tag{2.2}$$

after the application of N tests. Because only wrong fault-hypotheses are rejected, the set $\mathcal{H}(N)$ contains the true fault-hypothesis. In the best case, the cardinality of $\mathcal{H}(N)$ is one and the true fault in the plant is diagnosed.

Because service diagnosis is usually carried out in repair shops, the faults which are present can be assumed to be present at all times. Also, the set $\mathcal{F}_{\mathrm{CWA}}$ of all faults that may occur in the plant is assumed to be known. Without any additional information on the nature of the plant, the initial guess $\mathcal{H}(0)$ can be chosen as

$$\mathcal{H}(0) = \mathcal{H}_{\mathrm{CWA}}, \tag{2.3}$$

which guarantees the true fault-hypothesis to be in the initial guess.

The main idea of service diagnosis is the rejection of wrong fault-hypotheses. This is realized by *tests* T which determine an output hypothesis-set $\mathcal{H}_{\mathrm{Out}}$ from an input hypothesis-set $\mathcal{H}_{\mathrm{In}}$. Therefore, the application of a test is denoted by

$$\mathcal{H}_{\mathrm{Out}} = T \circ \mathcal{H}_{\mathrm{In}}. \tag{2.4}$$

The result $\mathcal{H}_{\mathrm{Out}}$ of the test T is obtained by removing the hypotheses in the set $\mathcal{H}_{\mathrm{Rej}}$ which become known to be wrong during the test from the input hypothesis-set $\mathcal{H}_{\mathrm{In}}$. A graphical interpretation of such a test is depicted in Fig. 2.2.

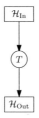

Figure 2.2: Application of a single test

More formally, a test T is a mapping from the power set of all possible fault-hypotheses $\mathcal{H}_{\mathrm{CWA}}$ to the power set of all possible fault-hypothesis $\mathcal{H}_{\mathrm{CWA}}$:

$$T : \quad 2^{\mathcal{H}_{\mathrm{CWA}}} \mapsto 2^{\mathcal{H}_{\mathrm{CWA}}}. \tag{2.5}$$

A test can only remove fault-hypotheses from a hypothesis-set. Therefore, the relationship

$$\mathcal{H}_{\mathrm{Out}} \subseteq \mathcal{H}_{\mathrm{In}} \tag{2.6}$$

between the input $\mathcal{H}_{\mathrm{In}}$ and the output $\mathcal{H}_{\mathrm{Out}}$ of a test holds. For this reason, the mapping T has the property

$$\mathcal{H} \supseteq T \circ \mathcal{H}. \tag{2.7}$$

Splitting up eqn. (2.7) into a proper subset and a set equality, one obtains two cases:

1. $\mathcal{H} \supset T \circ \mathcal{H}$: This corresponds to the case where the fault-hypotheses in the set $\mathcal{H}_{\mathrm{Rej}}$ become known to be wrong during the test. The diagnostic result *is refined* by removing these fault-hypotheses from the test input $\mathcal{H}_{\mathrm{In}}$ and the test result

$$\mathcal{H}_{\mathrm{Out}} = \mathcal{H}_{\mathrm{In}} \setminus \mathcal{H}_{\mathrm{Rej}}. \tag{2.8}$$

   is obtained. In this case, the test is said to be *successful*.

2. $\mathcal{H} = T \circ \mathcal{H}$: This corresponds to the case where no information on the system's fault-state could be obtained during the test. The diagnostic result $\mathcal{H}_{\mathrm{In}}$ *is not refined* and the test result

$$\mathcal{H}_{\mathrm{Out}} = \mathcal{H}_{\mathrm{In}}. \tag{2.9}$$

   is obtained. In this case, the test is said to be *unsuccessful*.

Each test can therefore be characterized by the set of fault-hypotheses $\mathcal{H}_{\text{Rej}}$ it may reject in the successful case. In summary, the result of a test is

$$\mathcal{H}_{\text{Out}} = T \circ \mathcal{H}_{\text{In}} = \left\{ \begin{array}{ll} \mathcal{H}_{\text{In}} \setminus \mathcal{H}_{\text{Rej}}, & \text{if the test is successful,} \\ \mathcal{H}_{\text{In}}, & \text{if the test is unsuccessful.} \end{array} \right. \qquad (2.10)$$

If the two possible results of a test are called $\mathcal{H}_{\text{P}}$ in the successful case and $\mathcal{H}_{\text{N}}$ in the unsuccessful case, a single test can be represented by Fig. 2.3.

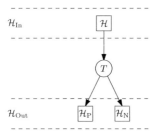

Figure 2.3: Possible results of a single diagnostic test

2.3 Sequential Refinement of the Diagnostic Result

Service diagnosis is a sequential search process in which the consecutive execution of tests refines the diagnostic result, eventually leading to a small hypothesis-set $\mathcal{H}(N)$.

Because the initial guess $\mathcal{H}(0)$ is in general a very large hypothesis-set, the execution of one single test does typically not directly result in a small hypothesis-set. However, the consecutive execution of several tests to the result of the previous test will, eventually, lead to the hypothesis-set $\mathcal{H}(N)$: Let the n-th test $T(n)$ applied to the result $\mathcal{H}(n)$ of the previous test yield the intermediate diagnostic result

$$\mathcal{H}(n + 1) = T(n) \circ \mathcal{H}(n). \qquad (2.11)$$

Then, one obtains the test-sequence

$$\mathcal{H}(N) = T(N - 1) \circ \cdots T(1) \circ T(0) \circ \mathcal{H}(0) \qquad (2.12)$$

which links the initial guess $\mathcal{H}(0)$ and the result of the sequential process of service diagnosis. This approach is illustrated in Fig. 2.4. Starting from the initial guess $\mathcal{H}(0)$, one of the diagnostic tests $T_1, ..., T_K$ is chosen to be the first test $T(0)$. This test results in the intermediate diagnostic result which is represented by the hypothesis-set $\mathcal{H}(1)$, which is either the refined hypothesis-set $\mathcal{H}_{\text{P}}$ or, if the test was unsuccessful, the set $\mathcal{H}_{\text{N}}$. The next test $T(1)$ is carried out on the hypothesis-set $\mathcal{H}(1)$ and so on. The choice of the test $T(n)$ in the n-th step is taken according to a diagnosis strategy. The diagnosis develops along the search tree depicted in Fig. 2.4, in this way resulting in a test-sequence.

This test-sequence entails a sequence of hypothesis-sets $\mathcal{H}(n)$, the cardinality of which becomes smaller with each successful test. In that way, the intermediate diagnostic result $\mathcal{H}(n)$ is refined, eventually leading to the hypothesis-set $\mathcal{H}(N)$ which is the result of the service diagnosis.

From this approach, two main questions arise:

1. Which test-sequences are advantageous for determining the true fault-hypothesis?

2. How can fast, inexpensive and reliable tests be constructed?

This thesis addresses the second question.

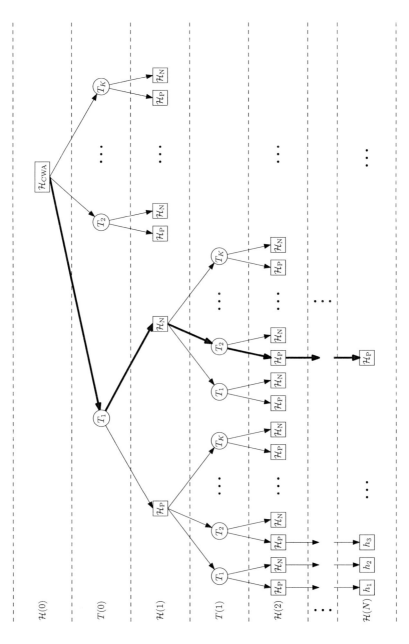

Figure 2.4: Sequential refinement of the diagnostic result

2.4   Consistency-Based Diagnosis

In this section, the principles of consistency-based diagnosis are introduced.

The plants which are considered in this thesis are mechatronic systems. These are dynamical systems which are subject to input signals $u(t)$ applied to the actuators of the plant. The plant answers to this excitation with output signals $y(t)$ which are provided by its sensors. The faults in the set $\mathcal{F}_{\text{CWA}}$ may alter the relation between the input signals $u(t)$ and the output signals $y(t)$, see Fig. 2.5.

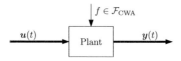

Figure 2.5: Plant

An approach to diagnosis, which does not require additional equipment, manual inspection, or the replacement of components can be based on this change. It uses the input signals $u(t)$ applied to the plant and the output signals $y(t)$ provided by the plant as depicted in Fig. 2.6 in order to infer on the fault-state of the plant. This approach is known from *process diagnosis*, where the main goal is to detect a fault $f(t)$ which is not present at all times, to distinguish it from perturbations $d(t)$, and to provide a set of fault candidates $\hat{\mathcal{F}}(t)$ in which the present fault is contained.

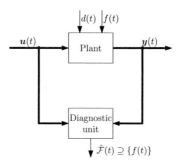

Figure 2.6: Process diagnosis

The main idea is to compare the signals $u(t)$ and $y(t)$ with the input signals and output signals that the plant exhibits in the case of a specific fault or in the fault-free case. This comparison is done in the diagnostic unit. Because it checks whether the observed signals are consistent with the behavior in a specific fault case or in the fault-free case, this kind of diagnosis is called *consistency-based* diagnosis. The behavior is usually provided by a model $M(f)$ of the plant which takes into account its fault-state f.

In process diagnosis, a fault f_i is called a fault candidate [82] if

$$y(t) = M(f_i) \circ u(t) \qquad (2.13)$$

holds. Then, the fault f_i is a plausible explanation for the output $y(t)$ under the excitation with the input signal $u(t)$. If

$$y(t) \neq M(f_i) \circ u(t) \qquad (2.14)$$

holds, f_i is not a fault candidate.

The link between the concept of fault candidates in process diagnosis and the concept of fault-hypotheses in service diagnosis is the following: If f_i is a fault candidate, the corresponding hypothesis h_i cannot be proven to be wrong. If

a)   the exclusion principle can be applied, and

b)   if there are fault candidates,

hypotheses on faults, which are not fault candidates, are wrong.

Equation (2.13) shows that a fault being a fault candidate may depend upon the input signal $u(t)$. In particular, for some input signals $u(t)$, the same output signals $y(t)$ may result in the case of the fault f_i as well as in the case of other faults. The same output signals $y(t)$ can even result if there is no fault at all. The input signal may therefore have an important impact when diagnosing dynamical systems.

This fact is used in *active diagnosis*, where specific input signals $a(t)$, often auxiliary to the original input signals, are applied to the plant. This principle is illustrated in Fig. 2.7.

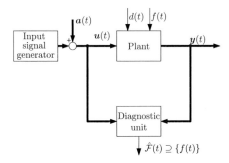

Figure 2.7: Active diagnosis

Methods for active diagnosis allow to avoid the problem that under a given operating strategy of a plant, some faults may never become a fault candidate although they are actually present. However, this approach generally has strong limitations, when used for process diagnosis, because the auxiliary signals $a(t)$ must be chosen in a way not to affect the primary system objectives like set-point tracking.

2.5 Automatic Testing as Active Diagnosis

Starting from the conclusion of the previous section, the utilization of dedicated test signals in automatic tests for service diagnosis is motivated.

The idea of process diagnosis to use the input and output signals of the plant to infer on its fault-state and the link between fault candidates and fault-hypotheses motivate the idea of *automatic tests*. These are tests which use the input and output signals of the plant for the rejection of wrong fault-hypotheses. The observation that the input signal $u(t)$ may have a strong impact on inferring whether a hypothesis is wrong leads to the approach to use specific input signals in automatic tests. In contrast to active diagnosis with small auxiliary signals [29, 130], where the primary system objectives have to be met, the input signals can be chosen arbitrarily in automatic tests. The input signals in automatic tests are called *test signals*. An automatic test

a) analyzes the input signals and the output signals of the plant, if

b) specific test signals are applied to the actuators of the plant.

In Fig. 2.8, the block scheme of such an automatic test is depicted. It consists of a diagnostic unit and a test signal generator which provides the test signal $u(t)$. The diagnostic unit realizes the rejection of wrong hypotheses from the set $\mathcal{H}(n)$ on the basis of the analysis of the input and output signals of the plant, thus providing the test result $\mathcal{H}(n+1)$. The analysis of the input and output signals in the diagnostic unit consists of checking whether the plant is a) in the desired operating region and b) a consistency test. The diagnostic unit also provides a signal $v(t)$ which influences the test signal generator.

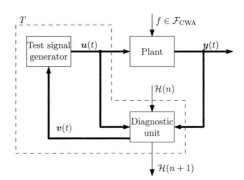

Figure 2.8: Automatic test

The main idea for the selection of the test signals is to highlight the effect of particular faults, whereas the effect of other faults is suppressed. The comparison of the input and output signals of the system can then be used for the rejection of fault-hypotheses. For example, it is assumed that the effect of a fault is suppressed, and the input and output signals of the system do not match the nominal behavior. Then, under the assumption that only one fault may be present at a time, the hypothesis that the suppressed fault is present, is wrong.

An example of this method can be illustrated with a simple mass-spring-damper-system as depicted in Fig. 2.9. The input $u(t)$ of the plant is the force $F(t)$ applied to the mass. The position of the mass $x(t)$ is the measured output $y(t)$ of the system. It is assumed that this plant can be subject to three different faults, each resulting in the deviation of one of the parameters c, d or m. Obviously, if the mass is not accelerated, a fault f_m which changes the parameter m will result in the same output signal as the fault-free plant. Therefore, if under this excitation there is a different output signal, a fault different from f_m is present. If only one fault is present, the hypothesis that the fault leading to the deviation of m is present, is wrong.

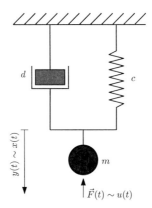

Figure 2.9: Illustrative example: mass-spring-damper system

In summary, checking whether the observed input and output signals match the nominal behavior of the plant is called consistency-based diagnosis. Inconsistency allows the rejection of fault-hypotheses. Analyzing whether under the excitation with a test signal a specific fault of a plant may lead to an output signal different from the one in the nominal case allows to infer on the diagnosability of the fault. Automatic tests exploit this property.

2.6   Literature Survey

The topic considered in this thesis overlaps with three fields: service diagnosis, active diagnosis, and structural methods, all of which are active areas of research. Therefore, this section is organized as follows: First, the idea of automatic tests is put in the context of service diagnosis. Then, approaches in the field of active diagnosis are discussed and compared to the approach to the test signal generation used in this thesis. An overview of the field of structural approaches in the context of diagnosis of dynamical systems is given. Finally, the works of other authors which combine structural models with active diagnosis are presented and compared to the approach taken in this thesis.

2.6.1 Service Diagnosis

A general introduction to the diagnosis of automotive systems can be found in [21]. The diagnosis of automotive systems is mainly driven by emission legislation and safety considerations which leads to requirements for the supervision of the system. However, the increasing complexity of the systems result in the necessity of efficient approaches to determine the fault in a system prior to its repair. While some efforts have been made to find advantageous test-sequences, [104], [127] and [128] thus answering the first question in Section 2.3, there are barely any publications which deal with finding automatic tests which themselves are reliable and inexpensive. This thesis aims at bridging this gap.

2.6.2 Active Diagnosis

Passive diagnosis approaches use the system's input and output signals to infer on the fault-state of the system, [61]. Dedicated excitation of the system allows to infer on the fault-state more precisely. Applying specific test signals to the system in order to detect and distinguish faults is called *active diagnosis*. Depending on whether the diagnosis shall be used during normal operation of the plant, these approaches can be divided into two groups:

a) The original functional goal of the system is preserved. The test signals are, therefore, not allowed to alter the behavior of the controlled system significantly, yet they have to highlight the fault enough to allow its detection.

b) The original functional goal of the system is abandoned. The test signals are therefore allowed to change the behavior of the controlled system significantly.

In the following, the literature is reviewed separately for the two groups.

a) Preservation of the original functional goal.

Although the problem solved by automatic tests clearly belongs to the second group, a short overview of the approaches which address the first problem is given in the following.

The constraint of preserving the original functional goal has been studied in a group of works that is concerned with determining small auxiliary inputs for diagnostic purposes. An early comprehensive work which considers this problem is [130].

Other early works on this topic originate from input design for identification. They are based on stochastic approaches and consider the discrimination of two models without aiming at fault detection in the first place. Examples are [57] and [123] in which the input is designed in a way to improve the Kullback-Leibler discrimination information, a measure for the difference between two stochastic behaviors. In [57], these approaches are used for the active detection of faults by distinguishing between two linear models. A dedicated approach to reduce the impact of the test signals on the functional goal was given in [126]. The approaches for fault detection are formulated in the frequency-domain in [58] and extended to fault isolation in [124]. A further extension to take into account model uncertainties can be found in [59] and [60]. In [125] a numerical example which shows that using the CUSUM algorithm for detection, a reduction of the mean-time to detection can be achieved if the above approaches are used.

A second approach to the problem of designing auxiliary signals is based on stochastic hypothesis testing which is done in [71] for linear noisy systems. It is used to distinguish between two hypotheses which corresponds to the problem of fault detection. The main idea is to choose input signals such that the mean time to false alarms is augmented and the average detection delay of the diagnosis is reduced. The approach is extended to the multi-hypotheses case which allows fault isolation in [72] and detailed in [67]. The methods are extended in [68] and [69] in order to distinguish between parameter vector changes of known magnitude and unknown direction using Fisher-information. In [66], a detailed overview of the case with the change from a known nominal mode to a known fault mode is given and [70] contains an extension towards testing hypotheses where neither the change of the magnitude nor the direction of the parameter vector is known.

In a third approach, particular interest has been directed to guaranteeing the detection of a fault in the presence of noise. In [91] this was done for systems with inequality bounded additive noise by applying auxiliary signals to the system to be diagnosed. This was extended to more general noise models in [98], to quadratically bounded perturbations in [92] and to uncertain models in [94], in order to deal with uncertain systems. Further extensions consider more general optimization criteria [47, 95] for the auxiliary input and nonzero initial conditions [93, 96]. All these approaches are summarized in [29]. A case study of the algorithms in [29] can be found in [31]. Detecting incipient faults instead of distinguishing different, but fixed fault scenarios with this approach is treated in [97] and [99]. Detecting faults in time-delayed systems is considered in [27, 28, 44]. In order to actively detect faults in sampled systems, [32, 45] take an approach based on piece-wise constant signals, which is detailed in [30] and summarized in [33]. Using linearization, the methods in [29] were applied in [26] to a nonlinear system. A more detailed description of this linearization-based approach can be found in [119]. A direct approach to the design of auxiliary signals for fault detection in nonlinear systems which does not make use of linearization, but uses optimization instead, is presented in [14] and detailed in [13]. The approaches in [29] were extended to the isolation of double faults in [48] and general multiple faults in [49]. More recently, the problem of detecting faults in discrete-time systems was studied, c.f. [16].

A fourth type of works considers detecting and isolating parametric faults in linear MIMO-systems. A particular decomposition allows to reformulate a controller which also yields an auxiliary input vector. The transfer function matrix from these inputs to the residuals is zero in the nominal case. Its entries are nonzero transfer functions in the faulty case. This matrix is called signature matrix in these works. In [87], the transfer functions in this matrix are used to determine a periodic signal which has a large impact on the residuals if applied to the auxiliary input and a small impact on the output of the plant. This main idea is extended to a scheme for fault-tolerant control in [84] using pre- and postfilters of the auxiliary input and the residuals. The approach is summed up in [86] and extended to uncertain systems in [85]. Using residual generators which are obtained from the model of the nominal plant and models of the plant subject to parameter-changes in [89], it is possible to capture changes in the working point of linearized models. Instead of evaluating the residuals with a threshold, in [100] a modified CUSUM-algorithm is used which exploits the gain and the phase-information present in the residual. This approach, which is summed up in [101] allows further isolation of parametric faults. In [88], gain and phase information of the signature matrix are exploited at a single frequency only, thus allowing an active diagnosis of a system with less detailed information on the system. An application of the described methods to a wind power turbine can be found in [90].

A fifth approach carries over concepts of sharing channels, which are known from communications engineering to the problem of detecting and isolating actuator faults in digital hydraulic systems [7].

b) Abandonment of the original functional goal.

The second group of works addresses the problem in which the original functional goal of the system is not required to be met. Then test signals which change the behavior of the system significantly can be applied to the system. In that way dropping a rather sharp constraint on the test signals, better fault detection and isolation properties are expectable. Because in this thesis the system to be diagnosed is assumed to have been brought to a repair shop, it is not necessary to maintain the original functional goal (noise, efficiency, emission goals etc.) of the system. The approaches developed in this thesis therefore belong to this group of works.

The problem of determining parameters in a system using nearly arbitrary input signals is well-known from system identification. Expressing a fault as an unknown parameter to be identified and designing the input signals with the tools known from system identification may solve the diagnostic problem if posed in that way. However, identifying the magnitude of a fault is not in the scope of this work and, therefore, the literature concerning input-signal design for parameter-identification is not considered.

The approach in [114] is based on the same decomposition as the one used in [87], but instead of applying test signals to an auxiliary input which do not change the system output significantly, a specific controller is used. This controller stabilizes the fault-free plant, but renders the faulty plant unstable. Switching to this controller and detecting the instability of the plant allows the detection of the fault. Further damage of the system is avoided by switching back to the nominal controller, once a fault is detected. The nominal controller is designed in a way that it stabilizes both, the fault-free and the faulty plant. A detailed description of the design of such controllers can be found in [115].

Another approach which is likewise based on a controller and does not allow to fulfill the original functional goal is described in [105]. It consists of masking the impact of a parametric fault in the system matrix of a linear system. A specific controller is used to steer the system such that the state-vector lives in the nullspace of the difference between the system matrix of the faulty and the fault-free system. The faulty and the fault-free system then exhibit the same behavior, the fault is hidden. By applying this controller, the residuals become zero and the fault associated to the specific controller is known to be present. Similarly, the idea to hide a fault is applied in [19] to isolate actuator faults in an unmanned aerial vehicle.

The principle of the design of test signals used in this thesis is similar to the approach of fault hiding. It is based on interrupting the cause-effect-chain between some faults and specific parts of the system-behavior. The faults of which the cause-effect-chain was interrupted cannot be the reason of an inconsistency of the input and output signals of the plant with the specific part of the system-behavior. In that way, hiding a fault in the complete behavior is a special case of the approach developed in this thesis. Also, the methods presented here, do not require the plant to be a linear system, thus accounting for the nature of mechatronic systems. In order to find test signals as well as the part of the behavior with which a consistency test is made, structural models are used in this thesis.

2.6.3 Structural Analysis for Diagnosis

The methods developed in this thesis are based on a special kind of model which is an abstraction of the constraints governing the plant's behavior. This model describes the couplings in the system with the help of a bipartite graph, the vertices of which represent the variables and the constraints which describe the system-behavior. The analysis of this graph, which is called *structure graph*, reveals diagnosability properties of the system and allows to determine relations to perform consistency tests such as residual generators. The methods for this analysis have been assembled to MATLAB toolboxes described for example in [23] and [54]. Structural analysis has been studied by the artificial intelligence community and the control engineering community with similar approaches which have been compared in [35], [36], [103] and [120]. This thesis takes the control engineering point of view and, therefore, in the following the AI approaches are not considered in detail.

Structure graph for FDI. An introduction to structural modeling and analysis can be found in [22] and the references therein. Because of the high level of abstraction, the structure graph is a useful tool for describing large-scale nonlinear systems. Typically, the structure graph is used to describe the fault-free system, faults are considered to be arbitrary violations of a model-constraint, although some references, c.f. [52], [63] and [80] use explicit fault models. In that way, only the relevant fault cases are considered which is similar to the approach of fault variables taken in this thesis. Properties which are important for the analysis of bipartite graphs have been investigated early in [46] without the scope of diagnosis. The first application of the structure graph to diagnosis in FDI was to determine constraint sets with the help of which it is possible to compute *analytical redundancy relations* (ARR) which allow a consistency test. The *matching*, which can likewise be found using the graph, allows to determine the order in which the equations in such a constraint set need to be solved and combined to obtain an ARR.

This was, for example, done in [34]. Finding an advantageous matching can also be used to take into account implementation aspects, c.f. [37] and [38].

In contrast to the above references which use a structure graph which describes the couplings in arbitrary operating regions, in this thesis a structure graph is used which holds only in specific operating regions.

MSOs. A central concept in structural analysis are minimal ARRs or minimal structurally overdetermined sets (MSO). These are the smallest subsets of equations of the system model which allow to determine a residual generator. A number of algorithms has been suggested to determine such sets, for example [56], [103], [121] and a group of works in [74], [75], [78], [79]. These eventually led to a particularly efficient algorithm in [76]. The properties and complexity of these algorithms are compared in [15]. The algorithm to determine MSOs used in this thesis is the basic algorithm from [76].

Invertibility. More recent publications consider the effect that the order in which the equations are solved for a variable defined by a matching may be infeasible. This is, for example, the case if equations are not invertible. In [41] and [42], the structure graph is modified to capture this effect, which is similar to the early approach in [62], [64] and [65]. The modification consists of ignoring such edges in the structure graph that represent the relation between a constraint and a variable that cannot be computed unambiguously from the other variables adjacent to this constraint. In this thesis, the same structural modification is used to capture the effect of non-invertibility which is for instance due to inequalities. This is a difference with respect to the approach to handling inequalities in [18].

Causality. Growing interest has recently also been directed to the question of causality which refers to whether an integrator or a differentiation w.r.t. time is used in order to express that a variable is determined from its derivative or vice versa. If only integrators are allowed, the term "integral causality" is used if only differentiation is allowed, the term "differential causality" is used. The term "mixed causality" is used if both relations are allowed. This has an impact on the generation of residuals, c.f. [117], the analysis of structural diagnosability, c.f. [17], [50], [51] and the sensor placement problem, c.f. [107]. However, in this thesis, differential causality is used.

Analysis of structural diagnosability properties. The algorithms in [76] allow to determine all possible constraint sets which can be used to determine residual generators. They therefore provide means to determine which faults may not be distinguished using these residual generators and threshold-checking. This property, which is important in passive diagnosis, is called structural isolability. An example of the structural analysis of diagnosability properties can be found in [40], where the DAMADICS-benchmark problem is analyzed. The analysis of the structural isolability also allows the selection of a residual generator out of a set of residual generators which have the same structural properties on the basis of further criteria. One possible criterion is the number of residuals required to reach desired fault isolability which is suggested in [118].

Sensor placement. The fact that usually not all faults are structurally isolable, has given rise to the question, where sensors need to be built in the system in order to achieve desired isolation properties. This problem is considered, for example, in [121] and [122], where the structural diagnosability properties are computed under the hypothesis that additional variables are measured directly. Whereas in these references the initial assumption is that all potential sensors are present and sensors are successively removed, the approach in [53] and [77] successively adds sensors to the system. A third approach to this problem is pursued in [83], [106],[108], [109], [110], [111], [112], [113] and [129], where optimization techniques are used to minimize a cost function which is associated to diagnosability properties.

However, instead of improving the diagnosability properties by additional sensors, this thesis aims at using only the sensors already built in the plant. The distinction between the faults is reached by the specific excitation of the actuators. In that way it becomes possible to distinguish between faults that are otherwise, i.e. structurally, not distinguishable.

2.6.4   Structural Approaches to Active Diagnosis

Structural approaches allow to determine residual generators for consistency tests in large-scale nonlinear systems. They may therefore realize an important step in the design of the diagnostic unit of an automatic test. However, since the structure graph known from the literature only describes which variables appear in which constraints, the approaches based on it do not capture the effect of specific system excitation. Only very few works deal with the fusion of structural approaches and active diagnosis.

The approach in [24] is concerned with isolating faults that are structurally not isolable by employing active diagnosis. It uses structural analysis of the nominal system to identify residual generators that contain system inputs in a first step. In a second step it is verified that the transfer from inputs to residuals is affected differently by faults on different constraints. A structural property which links the impact of faults on the system outputs to the structural condition in the second step is derived. Algorithms which allow to verify this property are presented in [55]. However, no proposition for the input signals is made in these references.

In [81], the combination of active diagnosis and structural methods has been used for a hybrid system, the structure of which depends upon its current discrete mode. The approach proposes to actively steer the system into a mode in which the continuous part of the system has a structure which is advantageous for fault detection and isolation. Therefore, the input signal used for diagnosis only concerns the discrete states of the system. A methodical approach to the problem that this steering may fail is not presented.

The approach in [39] studies the effect of adding and removing constraints in oder to capture the structural nature of different operating modes or systems which are subject to modifications. This reference is mainly concerned with reusing results of structural analysis which were obtained prior to a change in the set of constraints and not with active diagnosis. It is mentioned here because it is close to the concept of local structure graphs which results from a specific system excitation and which is introduced in this thesis.

In contrast to the references above, in this thesis a generalization of the structure graph is used which takes into account the influence of specific operating regions on the couplings between the variables. In particular, a method is developed to find operating regions in which some couplings between variables vanish.

The couplings between the variables if the system is restricted to such an operating region are described by a specific structure graph. This graph is called local structure graph. It results if edges that represent vanished couplings are dropped from the normal structure graph and constraints which describe the operating region are added. The local structure graph allows to determine so-called local residual generators. These are dynamical systems which provide signals which have the properties of residuals if the system is brought to an operating region in which the local structure graph holds. Also, a method to determine input signals which steer the system into the operating region in which a local structure graph holds is presented in this thesis.

In an earlier publication [2], the author starts from the idea that in specific operating regions a fault may not have any impact on the system-behavior. This allows to distinguish this fault from other faults which do not have this property. In the structure graph this is represented as follows: If the fault is represented by a variable in the structure graph and the system is brought to the specific operating region, the edge between the fault variable and the rest of the graph is dropped. In [2], an algorithm is given to identify operating regions in which edges between fault variables and constraints can be eliminated and an approach to steer the system into this operating region is proposed. This approach is used in [2] to distinguish faults in an SI engine. The result was validated with a more complex model of the engine in [43]. In [5], the method to determine the operating regions was presented in more detail at the example of a throttle valve.

In [3], the author generalizes the approach to interrupt the fault-effect-chain by choosing operating regions that allow to drop edges between constraints and variables other than fault variables in the structure graph. The local residual is introduced and a method is given to determine it from the structure graph which holds in the specific operating region. The main result in [3] is the detection of such an operating region by specific signals called validuals, which are likewise found by investigating the system structure. An approach to determine input signals which allow to steer the potentially faulty system is also developed in [3].

In [4], the constraints describing the operating region are incorporated in the structure graph which describes the couplings in this operating region, thus forming the local structure graph. A procedure to choose validuals and local residuals that are beneficial to fault isolation is given. This choice is based on the analysis of the sensitivity of a test consisting of validuals and local residuals. It allows to distinguish faults that are indistinguishable using standard structural analysis [22, 76, 77] or tests without the choice of advantageous operating regions and validuals [3]. The author details the approach in [6] with a process engineering example.

2.7 Assumptions and Problem Statement for This Thesis

In this section, an overview of the assumptions used in this thesis and a problem statement are given.

Single-Fault-Assumption. It is assumed that only one fault may be present in the plant at a time. The Single-Fault-Assumption is not a significant restriction in the case of service diagnosis for two reasons:

- If a plant exhibits an abnormal behavior due to a fault, service diagnosis is carried out as soon as possible in order to repair the plant. The probability of another fault occurring in the plant during this time is very unlikely.

- The occurrence of two or more faults at the same time can be considered as one single fault-state. Considering all combinations of faults results in an extremely large number. Although the methods developed in this thesis then become conservative, they remain valid.

Closed-World-Assumption. The second assumption in this thesis is that all faults that may occur in the plant are known prior to the design of a test. This assumption is referred to as the Closed-World-Assumption. Because the rejection of wrong fault-hypotheses is based on the exclusion principle which requires the totality of all hypotheses to be known, it is particularly important for the refinement of the diagnostic result. For two reasons, this assumption is less limiting then one would initially guess:

- In service diagnosis, typical faults which are, for example, due to aging or abrasion are known. This is due to a specific property of the development process used for most mechatronic systems and automotive systems in particular. In Section 3.3.3, this is explained in more detail.

- It is possible to include a fault which describes an unknown, unforeseen fault in the set of the considered faults $\mathcal{F}_{\text{CWA}}$.

Problem statement. The overall problem to be solved by service diagnosis (c.f. Section 2.1) can be reduced to determining automatic tests. This reduction is achieved by the stepwise solution of the diagnostic problem via a sequence of tests, each of which rejects wrong fault-hypotheses on the basis of the system's input and output signals. The reasoning that an automatic test consists of a test signal generator and a diagnostic unit allows to formulate the problem statement considered in this thesis as:

"Given a model of a plant, find a test signal generator and a diagnostic unit forming an automatic test which rejects wrong fault-hypotheses."

This problem statement motivates the following questions:

- Which kind of models should be used?

- What are the conditions for the existence of automatic tests?

- How can one construct a test signal generator?

- How can one construct a diagnostic unit?

- Under which conditions can a test reject wrong fault-hypotheses?

- Which are the fault-hypotheses $\mathcal{H}_{\mathrm{Rej}}$ a test may reject in that case?

2.8   Summary

In this chapter, service diagnosis was introduced as a sequential process which starts from a large set of fault-hypotheses with the goal to find the true fault-hypothesis. In this process, the consecutive execution of tests rejects wrong fault-hypotheses. In that way the intermediate diagnostic result is refined stepwise, thus solving the original problem.

From a description of the concepts of consistency-based diagnosis in Section 2.4, the idea of automatic tests was motivated. These are tests, which refine the intermediate diagnostic result on the basis of the input and output signals of the plant when excited with specific test signals.

Existing concepts and their differences with respect to the approaches in this thesis were reviewed in Section 2.6. The assumptions used in the thesis and a problem statement were given in Section 2.7.

Chapter 3

Modeling Framework

3.1 System-Behavior Representation by Constraint Sets

Consistency-based diagnosis requires a model of the behavior of the plant in order to check whether the observed input and output signals are consistent with this behavior. In this chapter, a way to illustrate the plant behavior and the modeling approach used in this thesis are presented.

In order to illustrate the behavior of a plant in the fault-free and in the faulty case, the notion of its behavior $\mathcal{B}$ is used. The behavior is the set of input-output pairs $(\boldsymbol{u}(t), \boldsymbol{y}(t))$ that are consistent with the plant. Figure 3.1 shows that the behavior $\mathcal{B}_0$ of the fault-free and the behavior $\mathcal{B}_f$ of the faulty plant are typically not the same. It is this property which allows diagnosis based on the input and output signals of the plant.

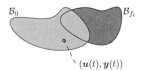

Figure 3.1: Behavior and I/O-pair

Although the illustration in Fig. 3.1 is useful for the explanation of the concepts presented in this thesis, it does not allow to determine a test signal and a diagnostic unit. For this purpose, an approach to model the plant behavior is introduced in this chapter. The dynamical systems considered are complex, continuous-time and nonlinear. Standard modeling approaches exhibit some inconveniences for the goal of consistency-based automatic tests for this system class: Linearization entails possibly (large) model errors when the system is taken to other operating points than the one used for linearization. Nonlinear state-space representations are difficult to handle for larger (complex) systems. Also, these modeling approaches consider the same degree of detail for all parts of a complex, large-scale system at the same time. They therefore result in large models

25

which are not suited for the model-based design of automatic tests. The main idea of the modeling approach used in this thesis is therefore to avoid the above problems by considering different degrees of detail at different steps of the analysis and the design of tests.

The first level of detail consists of a set of constraints which describes the entire system-behavior. The second level of detail uses an abstraction of this complete detailed model to determine specific subsets of the constraints. Only the detailed model-information contained in these constraint subsets is used to construct the test signal generator and the diagnostic unit of an automatic test. This realizes the complexity reduction during the design of a test.

The main contribution of this chapter is the introduction of the modeling approach of the first level of detail. Constraints c_i describe the couplings between the input and output signals of the plant via unknown intermediate variables. The constraints also contain a type of variables which represent the fault-state of the plant. Using variables which represent the system's fault-state like in [52, 63, 80] is an extension of the modeling framework in [76]. This extension will allow to determine the type of fault without determining its magnitude.

In this way, a set $\mathcal{C}$ of constraints c_i can be used to quantitatively describe a behavior as illustrated in Fig. 3.1. Since the cardinality of this constraint set is not restricted and the constraints themselves may be nonlinear, may contain differentation w.r.t. time, and may include inequalities and curves, the considered system class can be treated by such constraint set models.

The rest of the chapter is organized as follows: In Section 3.2, the considered system class is described and the illustration of the behavior is explained. The approach to modeling a system-behavior by a set of constraints is discussed in Section 3.3. An example which is used throughout this thesis to explain the new concepts is described in Section 3.4.

3.2   Behavioral Description of Dynamical Systems

In this section, the behavior of dynamical systems as the union of all input-output pairs allowed by the system is introduced.

The plants considered in this thesis are mechatronic systems. These systems consist of several components which allow the repair of the faulty plant by replacing the faulty component. The assumption that mechatronic systems are considered, leads to the following plant properties: The plants considered are:

- deterministic,

- continuous-time,

- nonlinear dynamical systems.

If a static SISO-system is subject to an input *value* $u \in \mathcal{V}_\mathcal{U}$ it provides an output *value* $y \in \mathcal{V}_y$. The set $\mathcal{V}_\mathcal{U} \subseteq \mathbb{R}$ is the set of all possible input values and the set $\mathcal{V}_y \subseteq \mathbb{R}$

is the set of all possible output values, respectively. The union of all pairs (u, y) which can be observed at this system forms a curve in the $\mathcal{V}_u\mathcal{V}_y$-plane. It describes the relation between input and output signals of the static SISO system completely. Obviously, this union may be different in different fault-states. In Fig. 3.2, this situation is depicted in the fault-free case and in the case of a fault.

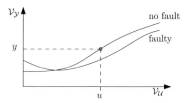

Figure 3.2: Behavior and I/O-pair of a static SISO-system

Similarly, if a dynamical system with the above properties is excited with a specific input *signal* $\boldsymbol{u}(t)$, it answers with a specific output *signal* $\boldsymbol{y}(t)$. The pair

$$(\boldsymbol{u}(t), \boldsymbol{y}(t)) \tag{3.1}$$

of this input signal and the corresponding output signal is called *input-output pair* or *I/O-pair*. The union of all I/O-pairs which can be observed at a dynamical system are called the *behavior* of the system. The behavior is denoted by $\mathcal{B}$. As for the static SISO system, a fault may change the behavior of the plant. It is therefore denoted by $\mathcal{B}_0$ in the fault-free case and by $\mathcal{B}_{f_i}$ in the case of fault f_i. Note that the behavior of a dynamical system depends upon its initial conditions as well. This effect is not considered in the following because the focus of this thesis is the impact of faults on the system-behavior and the initial conditions are assumed to be known.

Let $\mathcal{L}_\mathcal{U}$ be the set of all input signals $\boldsymbol{u}(t)$ and $\mathcal{L}_\mathcal{Y}$ the set of all output signals $\boldsymbol{y}(t)$. The behavior of a dynamical system can then be illustrated in the $\mathcal{L}_\mathcal{U}\mathcal{L}_\mathcal{Y}$-plane which is also called I/O-plane. In Fig. 3.3, this is done for the fault-free plant and the behavior in the case of fault f_i.

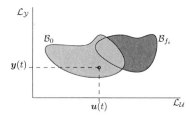

Figure 3.3: I/O-pair and behavior of a dynamical system in the I/O-plane

In this thesis, overlapping is used to indicate the intersection of two sets and hence behaviors that share the same I/O-pairs.

Limitations and benefits of the representation in the I/O-plane. The analogy between the representation of the mapping between two sets of scalar values (c.f. Fig. 3.2) and two sets of signals (c.f. Fig. 3.3) has two limitations: One limitation is that a set of scalar values like $\mathcal{V}_\mathcal{U}$ can be ordered which is not the case for a set of signals like $\mathcal{L}_\mathcal{U}$. To indicate this, no arrowheads at the axes $\mathcal{L}_\mathcal{U}$ and $\mathcal{L}_\mathcal{Y}$ are drawn in this work. The second limitation is, that the representation of a single behavior in the $\mathcal{L}_\mathcal{U}\mathcal{L}_\mathcal{Y}$-plane (c.f. Fig. 3.3) may lead to the impression that for one input signal, there is more than one output signal. This is in fact not the case since it contradicts the property of determinism in the considered system class. Therefore the I/O-plane has to be interpreted carefully in this regard.

However, throughout this work the I/O-plane with the axes $\mathcal{L}_\mathcal{U}$ and $\mathcal{L}_\mathcal{Y}$ will be used to explain behavior-related concepts. This is done for three reasons:

First, the analogy to the static SISO-case is kept. This facilitates the explanation of concepts that are based on specific system excitation.

Second, the explanation of most of the concepts requires the representation of both, signals and behaviors - that is sets of signal-pairs.

Third, many concepts are based on the relation between a specific input signal and the resulting output signal in the case of different faults. In this case it is useful to indicate the single input signal on the $\mathcal{L}_\mathcal{U}$-axis and the different output signals on the $\mathcal{L}_\mathcal{Y}$-axis.

3.3   Analytical Model of Dynamical Systems

In this section, the concept of analytical models is introduced.

An *analytical model* of a system is the description of its behavior by a set of variables $z_j \in \mathcal{Z}$ and a set of constraints $c_i \in \mathcal{C}$ relating to these variables. Both are detailed in the following.

3.3.1   System Variables

In this thesis, the term variable means a symbol in a mathematical expression. The value of this symbol can vary over time and, therefore, a variable obeys a signal. The variables occurring in the plant model are denoted by $z \in \mathcal{Z}$. The set $\mathcal{Z}$ of all variables can be partitioned into three different kinds of variables

$$\mathcal{Z} = \mathcal{X} \cup \mathcal{K} \cup \mathcal{F}, \tag{3.2}$$

the meaning of which is described in more detail in the following.

Known variables $\mathcal{K}$. The known variables $\mathcal{K}$ are those variables, which are applied to or measured at the plant directly: the inputs $u \in \mathcal{U}$ and the outputs $y \in \mathcal{Y}$ of the plant. The known variables are denoted by

$$\mathcal{K} = \mathcal{U} \cup \mathcal{Y}. \tag{3.3}$$

Unknown variables $\mathcal{X}$. The unknown variables $\mathcal{X}$ are those variables which are not applied to or measured at the plant directly and which do not describe the fault state of the plant. Examples of these variables are state-variables, physical quantities that are measured by a sensor or any other intermediate variable. They are denoted by

$$x \in \mathcal{X}. \tag{3.4}$$

Fault variables $\mathcal{F}$. The fault variables $\mathcal{F}$ represent the fault-state of the system. They are denoted by

$$f \in \mathcal{F}. \tag{3.5}$$

3.3.2 Constraint Types

In this thesis, constraints sets are used to describe behaviors. In the following, the notation

$$\mathcal{B}(\mathcal{C}) \tag{3.6}$$

is used to denote the behavior defined by the constraint set $\mathcal{C}$. Constraints $c_i \in \mathcal{C}$ describe the interaction of variables with each other. They are denoted in residual form by

$$c_i : \quad 0 = g_i(x_1, x_2, ..., u_1, u_2, ..., y_1, y_2, ..., f_1, f_2, ...) \tag{3.7}$$

with the scalar function $g_i(\cdot)$. In the following, different types of constraints are described.

Equations. An important type of constraints is the equation. In the analytical model of a mechatronic system, equations usually occur in the form of mass or energy balances and the equations which represent a sensor by linking unknown variables to known variables.

Curves and maps. Several effects in mechatronic systems are difficult to model using physical laws. This is, for example, the case with the impact of complex geometries on the flow resistance or friction effects. In that case, it is possible to describe the relation between the variables interfering in this physical effect by phenomenological models. Those relations are represented using curves or maps.

Differentiation. A special kind of constraints is used to describe the relation between a variable and its derivative w.r.t. time in dynamical systems:

$$c_i : \quad 0 = \dot{z}(t) - \frac{\mathrm{d}}{\mathrm{d}t} z(t). \tag{3.8}$$

This allows to give an analytical model which, apart from the differential constraints of the form eqn. (3.8), only consists of algebraic constraints.

Inequalities. Additionally to the equations and maps obtained from physical and phenomenological models, it is often possible to specify regions in which the value of some variables must lie. This can be represented by constraints of the form

$$c_i : \quad 0 < g_i(x_1, u_1, y_1, ...). \tag{3.9}$$

3.3.3   Qualitative Fault Models

The fault models used in this thesis describe different severities of a fault. This is done by the values of the fault variables f_i. By convention they are zero in the fault-free case and nonzero otherwise. With this convention, the hypothesis h_i that the fault f_i is present is

$$h_i : f_i \neq 0. \tag{3.10}$$

Different values of a fault variable f_i allow the description of different severities of this fault. In Fig. 3.4, the nominal behavior $\mathcal{B}_0$ and the behavior $\mathcal{B}_{f_i}$ in the case of fault f_i for different values of the fault variable are depicted.

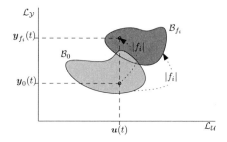

Figure 3.4: Behavior of qualitative fault models

The absolute value of the fault variable f_i moves the I/O-pairs in the I/O-plane. Typically, the larger the fault f_i, the larger the difference between the I/O-pairs forming $\mathcal{B}_0$ and $\mathcal{B}_{f_i}$. By definition, the fault variables f_i being zero entails the nominal behavior:

$$f_i = 0 \;\rightarrow\; \mathcal{B}_{f_i} = \mathcal{B}_0. \tag{3.11}$$

Different model-based diagnosis principles require different detail-levels of knowledge about the impact of possible faults. For example, approaches which are based on structural models like the one in [22], usually consider a fault to be an arbitrary violation of a constraint describing the plant behavior. The parity-space approach requires more detailed knowledge of the fault by modeling the fault as an additional signal which effects the plant. A third approach consists of modeling the fault-free and the faulty behavior of the plant by introducing parameters into the model which represent the fault. The values found for these parameters by parameter identification methods are then used to solve the diagnostic problem.

The idea of qualitative fault models used in this thesis is similar to the approach used in identification-based approaches. However, the diagnostic problem is not solved by determining the exact values of the fault variables. Instead, it is shown that specific fault variables are nonzero and, therefore, the other faults can be ruled out. This is a considerably simpler problem than the identification of exact parameters which may fail for a number of reasons. However, it is sufficient for the rejection of wrong fault hypotheses.

Typically, the more detailed the knowledge of the impact of a fault is, the better is the result of the diagnostic method. For some diagnostic problems, it is impossible to obtain detailed information on the impact of the faults prior to the design of the diagnostic function. Also, if not all possible faults are known, the Closed-World-Assumption may not hold. If faults or their impact on the behavior are unknown, it is not possible to use the fault-variable-approach.

However, a specific property of a common approach used in the development process of automotive systems ensures that all possible faults are known and their mechanism of action is understood. Therefore, the fault-variable-approach can be taken. A part of the development process of such systems is depicted in Fig. 3.5.

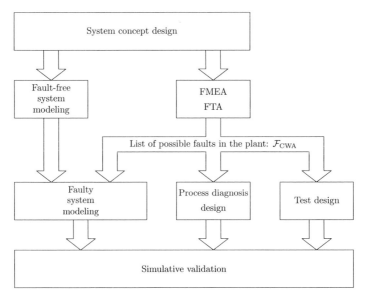

Figure 3.5: Common approach to test design ensuring the completeness of the considered faults

During the development of an automotive system, two analyses of the system under development are done: the FMEA (Failure Mode and Effects Analysis) and the FTA (Fault Tree Analysis), c.f. [20]. These methods allow to determine the faults which may occur in the system and the way they influence the system-behavior. The information obtained in that way is then used for two purposes: first, for the development of the diagnostic functions (e.g. process diagnosis and tests). Second, models of the plant subject to the relevant faults are used to simulate the plant behavior in the faulty case and to validate diagnostic functions.

Therefore, for automotive systems, one can safely assume that all faults and their mechanisms of action are known. For this reason, it is not only possible to use qualitative fault models but also to justify the rather sharp limitation of the Closed-World-Assumption.

3.3.4   Elimination-Minimal Representation

The analytical model of a plant is typically not a unique representation. Different arrangements of the constraints may describe the system-behavior correctly. The methods developed in this thesis are based on techniques for the elimination of unknown variables in a set of constraints. This requires the analytical model to obey a specific form:

The idea behind eliminating a variable from a constraint set is solving a constraint for this variable and injecting the result into the remaining constraints. Therefore, typically, one constraint is needed in order to eliminate a variable in another constraint.

However, if a variable only appears inside a particular expression in the set of constraints, it is also possible to solve a constraint for this particular expression instead of solving the constraint for the variable. The variable can then be eliminated by replacing the entire expression in the remaining constraints. This approach may also allow to eliminate more than one variable with one constraint: If more than one variable appear together only in a specific expression, it is possible to eliminate all these variables by solving for the expression and replacing the entire expression in the remaining constraints with the result.

If the elimination of more than one variable with one constraint is not possible, an analytical model is called *elimination-minimal*. In this thesis only elimination-minimal models are considered.

3.4   Running Example: Plant A

This section introduces a system which is used for various examples in this thesis.

Throughout this thesis, a simple first order system with an additional output will be used in order to illustrate the algorithms and the underlying concepts. The additional output is formed by a multiplication of the system's input with the output of the first order system. The block diagram of the plant is depicted in Fig. 3.6. It is assumed that this plant may be subject to five distinct faults, each influencing one of the parameters of the system, except f_5 which realizes an offset.

The constraints which govern the behavior of Plant A are given below.

$$
\mathcal{C} = \begin{cases}
c_1: & 0 = x_1(t) - (1+f_1)K_\mathrm{u}u(t) \\
c_2: & 0 = \dot{x}_2(t) + \frac{x_2(t)}{(1+f_2)T} - x_1(t) \\
c_3: & 0 = x_2(t) - \frac{1}{(1+f_3)K_{\mathrm{y1}}}y_1(t) \\
c_4: & 0 = x_3(t) - (1+f_4)K_\mathrm{M}x_2(t) \\
c_5: & 0 = y_2(t) - K_{\mathrm{y2}}x_1(t)x_3(t) - f_5 \\
d_1: & 0 = \dot{x}_2(t) - \frac{\mathrm{d}}{\mathrm{d}t}x_2(t)
\end{cases}
\tag{3.12}
$$

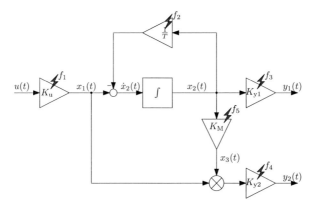

Figure 3.6: Running Example Plant A - block diagram

Example 3.1 (Analytical model of Plant A). *The analytical model of Plant A is*

$$
M_\text{A} \begin{cases}
\mathcal{C} &= \{c_1, c_2, c_3, c_4, c_5, d_1\} \\
\mathcal{K} &= \{u, y_1, y_2\} \\
\mathcal{X} &= \{x_1, x_2, \dot{x}_2, x_3\} \\
\mathcal{F} &= \{f_1, f_2, f_3, f_4, f_5\}.
\end{cases}
\tag{3.13}
$$

3.5 Summary

In this chapter, the class of systems considered in this thesis is described. The behavior of these systems is defined as the union of all I/O-pairs that may occur. The analytical model, a mathematical description of the behavior, is introduced. These models consist of a set of constraints which describe the relations between a set of variables, some of which are system inputs and system outputs, some of which are unknowns, and some of which represent the fault-state of the plant. A running example of an analytical model is given representing the behavior of a simple dynamical system in different fault-states.

Chapter 4

Diagnosability Analysis by Means of Directed Structure Graphs

4.1 Chapter Overview

In this chapter, questions important for the construction of the diagnostic unit in an automatic test are addressed. The main questions are:

1) How can one perform a consistency test in order to detect a fault or to distinguish between two faults?

2) Under which conditions is it possible to detect a fault on the basis of the behavior of the system? Under which conditions is it possible to distinguish between two faults?

In order to answer question 1), this chapter introduces the global residual $r(t)$ and a method to determine global residual generators which compute the signal $r(t)$ from the plant's input and output signals. The global residual allows to infer on the presence of a fault, if it is nonzero. This can be realized by a threshold check: $|r(t)| \geq \bar{r}$. The global residual has the property that, regardless of the excitation of the plant, it is zero in the fault-free case.

The property that it is possible to detect a fault in a plant on the basis of plant's behavior is called I/O-detectability. Similarly, two faults are called I/O-discriminable, if the behavior of the plant allows to distinguish between the two faults. This chapter answers question 2) in the sense of the relationship between the definition of these I/O-diagnosability properties and the analytical model of a plant.

An important result is given in Theorem 4.1 which states that a necessary and sufficient condition for the I/O-detectability of a fault is the existence of an input signal which entails different output signals in the fault-free and in the faulty case. A direct consequence of this theorem is that if a fault is I/O-detectable, there is a global residual generator. Since this residual generator can only detect a fault if the plant is excited with a specific input signal $\boldsymbol{u}(t)$, it is only of limited practical use. However, it motivates an approach to determine less conservative residual generators based on a representation of the couplings in the analytical model of the plant.

A second result is given in Theorem 4.5 stating that a fault is I/O-detectable if the input and output signals of a fault-free system contradict with a subset of the constraint set $\mathcal{C}_{f_i}$ which describes the behavior of the plant subject to fault f_i.

A third contribution is Theorem 4.6 which states that a fault is not I/O-detectable if the corresponding fault variable does not appear in a constraint set which can be contradicted. Similar results are obtained for the I/O-discriminability of a pair of faults.

The first question and the rather theoretical result that there is always a residual generator if a fault is I/O-detectable lead to the question: How can global residual generators be constructed and how is determined to which faults they are sensitive? In order to answer this, an extension of the structure graph in [76] is introduced as the structural representation of the plant. This bipartite graph represents the couplings among the variables in a constraint set by edges between two types of vertices which correspond to the constraints c_i and the variables z_j respectively, c.f. Fig. 4.1. Instead of representing a fault as an arbitrary violation of a constraint like in the larger part of the literature, similar to [52, 63, 80] the faults are represented by specific variable-vertices in this thesis. This extension and the incorporation of the information, which variable can be computed uniquely from which constraint lead to the directed global structure graph. An example of such a graph is depicted in Fig. 4.1.

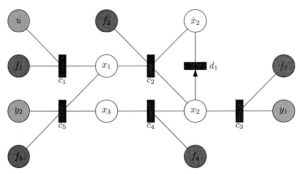

Figure 4.1: Directed global structure graph of Plant A

The structure graph is the second level of model-detail used in this thesis. With its help and algorithms from [76], it is possible to determine specific constraint sets, called globally minimal structurally overconstrained (GMSO). These constraint sets allow to determine global residual generators by eliminating all unknown variables from them. Using the global structure graph, it is possible to determine which fault variables do not interfere with a GMSO and, therefore, to which faults a corresponding residual generator is not sensitive. However, the analysis of the global structure graph does not allow to conclude that a global residual is sensitive to a fault if the fault variable appears in the GMSO used to obtain the residual generator. The properties of global structure graphs are considered in detail here because they are used in Chapter 5 for the analysis of the new local structure graphs.

The chapter closes with the important result that the conditions of the global structure graph which allow to compute a residual generator are neither necessary nor sufficient for the I/O-detectability of a fault. That is, using global structural analysis, one can not guarantee that a fault is I/O-detectable. However, typically, the global residual generators found with the analysis of the global structure graph are sensitive to the faults which appear in the corresponding GMSO.

4.2 Consistency Tests with Global Residual Generators

This section develops the idea of consistency tests and their realization by global residual generators.

The idea of consistency-based diagnosis is to check whether an I/O-pair observed at the plant's inputs and outputs is contained in the system's nominal behavior $\mathcal{B}_0$. Two possible situations may result: A) the I/O-pair observed at the plant *is* contained in the nominal behavior $\mathcal{B}_0$, and B) the I/O-pair observed at the plant *is not* contained in the nominal behavior $\mathcal{B}_0$. Both situations are illustrated in the I/O-plane in Fig. 4.2 with the I/O-pairs $(\boldsymbol{u}_\mathrm{A}(t), \boldsymbol{y}_\mathrm{A}(t))$ and $(\boldsymbol{u}_\mathrm{B}(t), \boldsymbol{y}_\mathrm{B}(t))$, respectively.

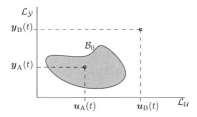

Figure 4.2: Principle of consistency tests

The relationship between those two situations and the fault-state of the plant is:

A) The observed I/O-pair is contained in the nominal behavior. No conclusion on the fault-state of the plant is possible since the measured I/O-pair may not only be contained in the nominal behavior, but in a faulty behavior as well. However, if the plant is not faulty, any I/O-pair observed at the plant must be contained in the nominal behavior. Therefore, an I/O-pair which is in the nominal behavior $\mathcal{B}_0$ is often understood as an indication that the plant is not faulty because it behaves like the fault-free plant.

B) The observed I/O-pair is not contained in the nominal behavior $\mathcal{B}_0$. The plant is then known to be faulty.

Since only the second case allows to infer on the fault-state of the plant, diagnostic techniques which are based on the plant behavior generally aim at detecting that an observed I/O-pair is not contained in the nominal behavior $\mathcal{B}_0$ – or in other words is inconsistent with the nominal behavior. The above reasoning motivates the definition of *global residuals* a specific kind of signals which match the properties A) and B) to the comparison of a signal value with zero. The following definition is taken from the definition of residuals in [22] and adapted to the models used here:

Definition 4.1 (Global residual). *The signal $r(t)$ is called* global residual *if it has the properties*

$$\begin{aligned} r(t) \neq 0 &\quad\Rightarrow\quad \exists\; f_i \neq 0 \\ f_i = 0\; \forall\; i &\quad\Rightarrow\quad r(t) = 0\; \forall\; t. \end{aligned} \tag{4.1}$$

A dynamical system which computes a global residual from the input signals and the output signals of the plant is called global residual generator *and denoted by*

$$r(t) = r\left(\boldsymbol{u}(t), \boldsymbol{y}(t), \frac{\mathrm{d}}{\mathrm{d}t}\right). \tag{4.2}$$

In this context, the notation $\frac{\mathrm{d}}{\mathrm{d}t}$ signifies that also derivatives w.r.t. time of any order of the input signals $\boldsymbol{u}(t)$ and the output signals $\boldsymbol{y}(t)$ may occur in the residual generator. Computing a global residual and checking whether it is zero allows to decide whether the I/O-pair observed at the plant is inconsistent with the nominal behavior. Note that for global residuals, the above properties hold for all t, regardless of the plant's excitation $\boldsymbol{u}(t)$ or answer $\boldsymbol{y}(t)$. A block diagram of the application of a global residual generator is depicted in Fig. 4.3.

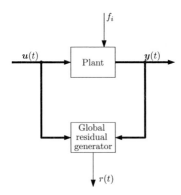

Figure 4.3: Residual generator realizing a consistency test

In the literature, the computation of a residual and its evaluation by threshold checking are often considered separately. However, evaluating a residual with a threshold check and checking whether a residual that was treated by a dead zone is zero are the same. Considering the original definition of a (global) residual allows to check whether the residual is nonzero. Because this often facilitates the reasoning, it is used throughout this thesis. The following reasoning justifies this approach: If $r(t)$ is a global residual, the output

$$\tilde{r}(t) = g(r(t)) \tag{4.3}$$

of any scalar function $g(\cdot)$ with the properties $g(0) = 0$ and $\exists\, x : g(x) \neq 0$ is a global residual as well. This is because if $r(t) = 0 \; \forall\; t$, also $\tilde{r}(t) = 0 \; \forall\; t$ holds true, and if $\exists\, t : \tilde{r}(t) \neq 0$, also $\exists\, t : r(t) \neq 0$ holds true. This is, for example, the case for functions $g(\cdot)$ which realize thresholds by dead zones.

The original condition of residuals for a consistency check can be applied to the modified residual $\tilde{r}(t)$. This leads to the check

$$\tilde{r}(t) = g(r(t)) \overset{?}{\neq} 0, \tag{4.4}$$

which can be used to achieve robustness with repect to imperfect models or perturbations. The same reasoning applies to any residual $r(t)$ filtered with a low-pass filter $G(s)$ with zero initial conditions. With the Laplace-transformation $\mathcal{L}$ and its inverse $\mathcal{L}^{-1}$, the signal

$$\tilde{r}(t) = \mathcal{L}^{-1}\left(G(s) \cdot \mathcal{L}(r(t))\right) \tag{4.5}$$

is a residual, because if $r(t) = 0 \; \forall\; t$, also $\tilde{r}(t) = 0 \; \forall\; t$ holds true and if $\exists\, t : \tilde{r}(t) \neq 0$, assuming the initial conditions to be zero, $\exists\, t : r(t) \neq 0$ holds true. Therefore, any low-pass filter can be used to achieve robustness in the presence of noise.

Obviously, functions and filters with the above properties conserve the properties of residuals, although they do not necessarily have the same results for the same I/O-pairs. The definition of a global residual generator therefore does not require $r(t) = 0$ to be equivalent with the consistency of an observed I/O-pair with the nominal behavior of the plant. The design of the set of I/O-pairs with which the global residual generator checks an I/O-pair for consistency is a trade-off between robustness due to uncertainties, noise, perturbation on one side and missed faults on the other side.

The difference between the nominal behavior $\mathcal{B}_0$ of a plant and the set of I/O-pairs which lead to a zero global residual is illustrated in the I/O-plane in Fig. 4.4.

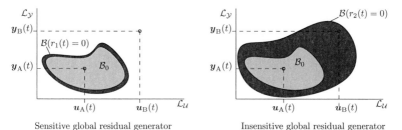

Sensitive global residual generator Insensitive global residual generator

Figure 4.4: Consistency tests realized by a sensitive global residual generator $r_1(\boldsymbol{u}, \boldsymbol{y}, \frac{\mathrm{d}}{\mathrm{d}t})$ and an insensitive global residual generator $r_2(\boldsymbol{u}, \boldsymbol{y}, \frac{\mathrm{d}}{\mathrm{d}t})$

Here, the expression $\mathcal{B}(r_i(t) = 0)$ is used as a way to describe the set of I/O-pairs which lead to the global residual $r_i(t)$ being zero.

The residual $r_1(t)$ is determined by a sensitive global residual generator, the residual $r_2(t)$ by an insensitive one. The behavior of both of the global residual generators, the sensitive and the insensitive one, contain the nominal behavior of the plant. Therefore, an arbitrary I/O-pair $(\boldsymbol{u}_A(t), \boldsymbol{y}_A(t)) \in \mathcal{B}_0$ results in the global residual generators indicating the consistency of the I/O-pair with the nominal behavior.[1]

The difference between the behavior which a sensitive global residual generator checks consistency of I/O-pairs with and the nominal behavior $\mathcal{B}_0$ of the plant is small. For this reason, the sensitive global residual $r_1(t)$ indicates inconsistency of the I/O-pair $(\boldsymbol{u}_B(t), \boldsymbol{y}_B(t))$ with $\mathcal{B}(r_1(t) = 0)$ – which is also true for the nominal behavior of the plant. The residual $r_1(t)$ allows to detect a fault which entails $(\boldsymbol{u}_B(t), \boldsymbol{y}_B(t))$ successfully. Since the I/O-pair $(\boldsymbol{u}_B(t), \boldsymbol{y}_B(t))$ is in the behavior the insensitive residual $r_2(t)$ checks consistency with, its result is $r_2(t) = 0$. This is the case although the I/O-pair $(\boldsymbol{u}_B(t), \boldsymbol{y}_B(t))$ is inconsistent with the nominal behavior of the plant – the fault is missed. This is in accordance with the definition of a global residual and one of the reasons, why from a zero residual, one cannot conclude on the fault-freeness of the plant.

Although the focus of this thesis is not the trade-off between missed faults and false detection, the above reasoning will be of use in the analysis of the relationship between structural and I/O-diagnosability properties.

Often, it is possible to derive more than one residual generator for a given plant. In that case, it is typically possible to determine residual generators that are insensitive to I/O-pairs which result from specific faults – although the I/O-pairs are not contained in the nominal behavior. It is this property that can be the key to discriminating different faults: Let $r_1\left(\boldsymbol{u}, \boldsymbol{y}, \frac{\mathrm{d}}{\mathrm{d}t}\right)$ and $r_2\left(\boldsymbol{u}, \boldsymbol{y}, \frac{\mathrm{d}}{\mathrm{d}t}\right)$ be two residual generators. Furthermore, let the I/O-pairs measured at a plant with fault f_2 never lead to a nonzero residual $r_1(t)$ and the I/O-pairs measured at a plant which is subject to fault f_1 never lead to a nonzero residual $r_2(t)$. It is then possible to discriminate the two faults by checking, which residual is zero and a Boolean decision logic.

An interpretation of this approach in the I/O-plane is illustrated in Fig. 4.5. The corresponding block-diagramm is depicted in Fig. 4.6. The behaviors $\mathcal{B}(r_i(t) = 0)$ contain the nominal behavior of the plant. Therefore, the signals $r_i(t)$ are global residuals. However, if the I/O-pair $(\boldsymbol{u}_B(t), \boldsymbol{y}_B(t))$ is observed at the plant, the global residuals $r_1(t)$ and $r_2(t)$ are nonzero while the global residual $r_3(t)$ is zero. An intelligent design of the residual generators $r_i\left(\boldsymbol{u}, \boldsymbol{y}, \frac{\mathrm{d}}{\mathrm{d}t}\right)$ and thus of the corresponding behaviors $\mathcal{B}(r_i(t) = 0)$ may then allow to distinguish between different faults. Note that in order to not miss out on any fault, the intersection of the behaviors $\mathcal{B}(r_i(t) = 0)$ should be the nominal behavior $\mathcal{B}_0$ of the plant.

However, the concept of fault detection and discrimination with the help of global residual generators entails two important questions:

• What are the conditions of a plant and its behavior that allow the detection and discrimination of faults on the basis of its input and output signals?

---

[1]Otherwise, the signals $r_1(t)$ and $r_2(t)$ would have violated the property $f_i = 0 \Rightarrow r(t) = 0$, because $f_i = 0$ entails $\mathcal{B}_0$ and, therefore, $r_1(t)$ and $r_2(t)$ would not be global residual generators.

- How can one construct global residual generators and determine which faults may not have any impact on the value of the global residual?

In order to answer these questions, the following four sections are concerned with the analysis of diagnosability properties. In Section 4.7 an approach to the design of global residual generators is presented, and in Section 4.8 the relationship between properties concerning the analytical model of the plant and diagnosability properties of the plant's faults are analyzed.

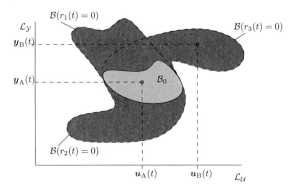

Figure 4.5: Consistency tests realized by a bank of residual generators

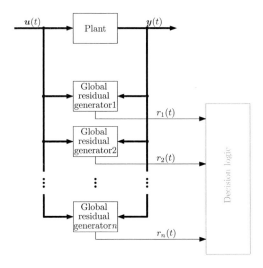

Figure 4.6: Bank of residual generators for fault discrimination

4.3   I/O-Detectability

This section discusses the concept of I/O-detectability, which answers the question whether it is possible to determine the presence of a fault in a plant on the basis of the plant's I/O-behavior. Conditions for the analytical model which allow to determine whether a specific fault is I/O-detectable are derived. In the following, it is assumed that the initial conditions are known.

A fault f_i is I/O-detectable, if it is possible to infer on the presence of this fault on the basis of the input signals and output signals of the plant. This is the case, if the fault changes the behavior of the plant which leads to the following definition:

Definition 4.2 (I/O-detectability). *A fault f_i is called* I/O-detectable, *if the behavior $\mathcal{B}_{f_i}$ is different from the nominal behavior $\mathcal{B}_0$ of the plant:*

$$\mathcal{B}_0 \neq \mathcal{B}_{f_i}, \ \forall \ f_i \neq 0. \tag{4.6}$$

The typical behavior of a plant in which the fault f_i is I/O-detectable according to this definition is depicted in Fig. 4.7.

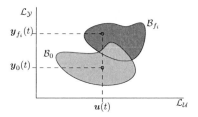

Figure 4.7: I/O-detectable fault f_i

Definition 4.2 is satisfied, if there is at least one I/O-pair, in which the faulty behavior and the nominal behavior are different:

Lemma 4.1 (I/O-detectability). *A fault f_i is* I/O-detectable, *if and only if it may entail an I/O-pair which is not contained in the nominal behavior $\mathcal{B}_0$:*

$$\exists \ \left(\boldsymbol{u}(t), \boldsymbol{y}_{f_i}(t) \right) \notin \mathcal{B}_0, \ \forall \ f_i \neq 0 \tag{4.7}$$

where $\boldsymbol{y}_{f_i}(t)$ is the output signal of the faulty plant, if excited with the input signal $\boldsymbol{u}(t)$.

Proof. Let $\boldsymbol{u}(t)$ be the input signal with which the plant is excited. In the case of fault f_i, that is $f_i \neq 0$, the plant answers with $\boldsymbol{y}_{f_i}(t)$. The pair $\left(\boldsymbol{u}(t), \boldsymbol{y}_{f_i}(t)\right)$ is contained in the behavior $\mathcal{B}_{f_i}$ of the faulty plant because the definition

$$\left(\boldsymbol{u}(t), \boldsymbol{y}_{f_i}(t)\right) \in \mathcal{B}_{f_i}, \ \forall \ f_i \neq 0 \tag{4.8}$$

always holds true. If the I/O-pair is not contained in the fault-free behavior $\mathcal{B}_0$,

$$\left(\boldsymbol{u}(t), \boldsymbol{y}_{f_i}(t)\right) \notin \mathcal{B}_0, \ \forall \ f_i \neq 0 \tag{4.9}$$

holds true. If that is the case, there is an element which is in the set $\mathcal{B}_{f_i}$, but not in the set $\mathcal{B}_0$ for all fault-magnitudes other than zero. Because two sets are distinct, if there is at least one element in one of the sets that is not contained in the other set, the sets $\mathcal{B}_0$ and $\mathcal{B}_{f_i}$, $\forall \ f_i \neq 0$ are distinct which is the same as Definition 4.2. $\qquad\square$

The practical meaning of Lemma 4.1 is that a fault is I/O-detectable if there is an input signal which results in a different output signal in the fault-free case than in the faulty case. This will also lead to Theorem 4.1.

The property described in Lemma 4.1 is used in consistency-based process diagnosis: It is checked whether the current I/O-pair of the plant is contained in, e.g. consistent with the nominal plant behavior $\mathcal{B}_0$. If this is not the case, it is inferred on the presence of a fault. In Fig. 4.7, this is illustrated with the pair $\left(\boldsymbol{u}(t), \boldsymbol{y}_{f_i}(t)\right)$ which is not contained in the nominal behavior $\mathcal{B}_0$.

Instead of describing a fault being I/O-detectable if it entails an I/O-pair that is not contained in the nominal behavior, it is also possible to check whether there is an I/O-pair observed at a fault-free plant which is not contained in the faulty behavior $\mathcal{B}_{f_i}$. In Fig. 4.7, this is illustrated with the I/O-pair $(\boldsymbol{u}(t), \boldsymbol{y}_0(t))$, which is not contained in the behavior $\mathcal{B}_{f_i}$. This approach also satisfies Definition 4.2, which leads to Lemma 4.2. While Lemma 4.2 is less intuitive than Lemma 4.1, it facilitates the reasoning on structural detectability of a fault later on.

Lemma 4.2 (I/O-detectability). *A fault f_i is I/O-detectable, if and only if the fault-free plant may entail an I/O-pair which is not contained in the faulty behavior $\mathcal{B}_{f_i}$:*

$$\exists \ (\boldsymbol{u}(t), \boldsymbol{y}_0(t)) \notin \mathcal{B}_{f_i}, \ \forall \ f_i \neq 0 \tag{4.10}$$

where $\boldsymbol{y}_0(t)$ is the output signal of the fault-free plant, if excited with the input signal $\boldsymbol{u}(t)$.

Proof. The proof is similar to the one for Lemma 4.1. By definition,

$$(\boldsymbol{u}(t), \boldsymbol{y}_0(t)) \in \mathcal{B}_0 \tag{4.11}$$

holds. If there is an I/O-pair, for which

$$(\boldsymbol{u}(t), \boldsymbol{y}_0(t)) \notin \mathcal{B}_{f_i}, \ \forall \ f_i \neq 0 \tag{4.12}$$

holds true, it follows that

$$\mathcal{B}_0 \neq \mathcal{B}_{f_i}, \ \forall \ f_i \neq 0 \tag{4.13}$$

holds, which is Definition 4.2.   □

The concept of the behavior as a set of I/O-pairs is useful for fundamental reasoning on the I/O-diagnosability and its illustration. However, it is only of limited use for the analysis of the analytical model of a given plant. In the following, the reasoning is therefore extended to the constraint set $\mathcal{C}$ in an analytical model which describes the plant's behavior.

If all fault variables in an analytical model are set to zero, it describes the nominal behavior $\mathcal{B}_0$ of the plant. The index zero is used to indicate that this was done:

$$\mathcal{B}_0 = \mathcal{B}(\mathcal{C}_0). \tag{4.14}$$

In the same way, the index f_i is used to indicate that all fault variables except f_i are set to zero and, therefore, the constraint set $\mathcal{C}_{f_i}$ describes the faulty behavior $\mathcal{B}_{f_i}$:

$$\mathcal{B}_{f_i} = \mathcal{B}(\mathcal{C}_{f_i}) \tag{4.15}$$

If this notation is used and if the output signal $\boldsymbol{y}(t)$ resulting from the excitation of a plant described by the set of constraints $\mathcal{C}$ with the input signal $\boldsymbol{u}(t)$ is denoted by

$$\boldsymbol{y}(t) = \mathcal{C} \circ \boldsymbol{u}(t), \tag{4.16}$$

the following theorem for I/O-detectability holds true:

Theorem 4.1 (I/O-detectability). *A fault f_i is I/O-detectable if and only if there is an input signal $\boldsymbol{u}(t)$ so that the output signal*

$$\boldsymbol{y}_{f_i}(t) = \mathcal{C}_{f_i} \circ \boldsymbol{u}(t) \tag{4.17}$$

of the faulty plant is not the same as the output signal

$$\boldsymbol{y}_0(t) = \mathcal{C}_0 \circ \boldsymbol{u}(t) \tag{4.18}$$

of the fault-free plant: $\boldsymbol{y}_{f_i} \neq \boldsymbol{y}_0, \ \forall \ f_i \neq 0.$

The following proof uses the fact that the considered systems are deterministic systems with known initial conditions.

Proof. In deterministic systems with known initial conditions, for every input signal $\boldsymbol{u}(t)$ there is exactly one output signal $\boldsymbol{y}(t) = \mathcal{C} \circ \boldsymbol{u}(t)$. For that reason, the behavior $\mathcal{B}_0$ does not contain I/O-pairs with the same input signal, but different output signals:

$$\boldsymbol{y}_1(t) = \mathcal{C}_0 \circ \boldsymbol{u}_1(t), \quad \boldsymbol{y}_2(t) = \mathcal{C}_0 \circ \boldsymbol{u}_2(t) \tag{4.19}$$

$$\boldsymbol{u}_1(t) = \boldsymbol{u}_2(t) \Rightarrow \boldsymbol{y}_1(t) = \boldsymbol{y}_2(t). \tag{4.20}$$

Because by definition

$$(\boldsymbol{u}(t), \boldsymbol{y}_0(t)) \in \mathcal{B}_0 \tag{4.21}$$

holds for the fault-free plant, together with the above property of deterministic systems,

$$\exists \ \left(\boldsymbol{u}(t), \boldsymbol{y}_{f_i}(t)\right) \notin \mathcal{B}_0, \ \forall \ f_i \neq 0 \tag{4.22}$$

must hold, if

$$\exists \ \boldsymbol{u}(t) : \boldsymbol{y}_{f_i}(t) \neq \boldsymbol{y}_0(t), \ \forall \ f_i \neq 0 \tag{4.23}$$

holds true. With the model of the fault-free plant

$$\boldsymbol{y}_0(t) = \mathcal{C}_0 \circ \boldsymbol{u}(t) \tag{4.24}$$

and the model of the faulty system

$$\boldsymbol{y}_{f_i}(t) = \mathcal{C}_{f_i} \circ \boldsymbol{u}(t), \ \forall \ f_i \neq 0 \tag{4.25}$$

eqn. (4.23) can be expressed as

$$\exists \ \boldsymbol{u}(t) : \mathcal{C}_0 \circ \boldsymbol{u}(t) \neq \mathcal{C}_{f_i} \circ \boldsymbol{u}(t), \ \forall \ f_i \neq 0. \tag{4.26}$$

If eqn. (4.26) holds, eqn. (4.22) holds as well, which corresponds to Lemma 4.1. □

The relationship between consistency tests with the help of global residual generators and the above theorem is given in the following remark.

Remark 4.1. *Lemma 4.1 allows to conclude that if a fault f_i is I/O-detectable, it is always possible to construct a global residual generator because a signal of the form*

$$r_1(t) = \begin{cases} 1, & \text{if } \boldsymbol{y}(t) = \boldsymbol{y}_\mathrm{B}(t) \ \text{for a specific } \boldsymbol{u}_\mathrm{A}(t) \\ 0, & \text{otherwise} \end{cases} \tag{4.27}$$

satisfies the properties given in Definition 4.1. Such a global residual generator infers on the presence of a fault, if, under the excitation with a specifc input signal, there is a specific output signal. In that way, only one I/O-pair in the I/O-plane is used for diagnosis, as depicted in Fig. 4.8. Theorem 4.1 allows the conclusion that even

$$r_2(t) = \begin{cases} 1, & \text{if } \boldsymbol{y}(t) \neq \boldsymbol{y}_\mathrm{A}(t) \ \text{for a specific } \boldsymbol{u}_\mathrm{A}(t) \\ 0, & \text{otherwise} \end{cases} \tag{4.28}$$

is a global residual generator which is more general than eqn. (4.27). However, such residual generators are of very limited applicability: When used for process diagnosis, it is very rare that the exact $\boldsymbol{u}_\mathrm{A}(t)$ can be observed at the plant during its operation. This could be overcome in service diagnosis, where the exact $\boldsymbol{u}_\mathrm{A}(t)$ can be applied to the plant, but then for a given residual it is difficult to conclude, which fault is present. Nevertheless, such residual generators are often used in end-of-line-tests in production where the main goal is fault detection.

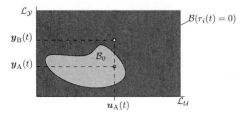

Figure 4.8: Guarantee for the existence of a global residual generator

In the following, the notation $\mathcal{C}((\boldsymbol{u}(t), \boldsymbol{y}(t)))$ is used to describe that the signals $\boldsymbol{u}(t)$ and $\boldsymbol{y}(t)$ are injected into the constraint set $\mathcal{C}$ for the known variables $\mathcal{U}$ and $\mathcal{Y}$ respectively. Then, the following necessary and sufficient condition for the I/O-detectability of a fault holds. It follows the idea of Lemma 4.2 to investigate whether an I/O-pair resulting from the fault-free plant is contained in the faulty behavior.

Theorem 4.2 (Necessary and sufficient condition for I/O-detectability).
A fault f_i is I/O-detectable if and only if there is an input signal $\boldsymbol{u}(t)$ such that the constraint set

$$\mathcal{C}_{f_i}((\boldsymbol{u}(t), \boldsymbol{y}_0(t))) = \mathcal{C}_{f_i}((\boldsymbol{u}(t), \mathcal{C}_0 \circ \boldsymbol{u}(t))), \quad f_i \neq 0 \tag{4.29}$$

does not have a solution for the unknown variables $\mathcal{X}$, if $\boldsymbol{y}_0(t)$ is the output signal, with which the fault-free plant answers to the input signal $\boldsymbol{u}(t)$.

Proof. With Theorem 4.1, the above theorem is equivalent to the statement

$$\exists\, \boldsymbol{u}(t) : \boldsymbol{y}_0(t) \neq \boldsymbol{y}_{f_i}(t) \Leftrightarrow \exists\, \boldsymbol{u}(t) : \mathcal{C}_{f_i}\left((\boldsymbol{u}(t), \boldsymbol{y}_0(t))\right) \quad \begin{array}{l}\text{has NO solution for the}\\ \text{unknown variables in it.}\end{array} \tag{4.30}$$

The following proof consists of treating necessity and sufficiency in two different steps. The first step consists of proving that from

$$\exists\, \boldsymbol{u}(t) : \mathcal{C}_{f_i}\left((\boldsymbol{u}(t), \boldsymbol{y}_0(t))\right) \quad \begin{array}{l}\text{has NO solution for the}\\ \text{unknown variables in it}\end{array} \tag{4.31}$$

the I/O-detectability of f_i follows. By definition,

$$\mathcal{C}_{f_i}\left(\left(\boldsymbol{u}(t), \boldsymbol{y}_{f_i}(t)\right)\right) \tag{4.32}$$

has at least one solution for the unknown variables in it. Therefore,

$$\boldsymbol{y}_0(t) = \boldsymbol{y}_{f_i}(t) \,\forall\, \boldsymbol{u}(t) \Rightarrow \mathcal{C}_{f_i}\left((\boldsymbol{u}(t), \boldsymbol{y}_0(t))\right) \quad \begin{array}{l}\text{has at least one solution for the}\\ \text{unknown variables in it } \forall\, \boldsymbol{u}(t)\end{array} \tag{4.33}$$

holds. Inversion of the above statement yields

$$\exists\, \boldsymbol{u}(t) : \boldsymbol{y}_0(t) \neq \boldsymbol{y}_{f_i}(t) \Leftarrow \exists\, \boldsymbol{u}(t) : \mathcal{C}_{f_i}\left((\boldsymbol{u}(t), \boldsymbol{y}_0(t))\right) \quad \begin{array}{l} \text{has NO solution for the} \\ \text{unknown variables in it.} \end{array} \quad (4.34)$$

The left part is equal to the I/O-detectability condition in Theorem 4.1 and, therefore, from

$$\exists\, \boldsymbol{u}(t) : \mathcal{C}_{f_i}\left((\boldsymbol{u}(t), \boldsymbol{y}_0(t))\right) \quad \begin{array}{l} \text{has NO solution for the} \\ \text{unknown variables in it} \end{array} \quad (4.35)$$

the I/O-detectability of fault f_i follows.

The second step consists of proving that from $\exists\, \boldsymbol{u}(t) : \boldsymbol{y}_0(t) \neq \boldsymbol{y}_{f_i}(t)$ the statement

$$\exists\, \boldsymbol{u}(t) : \mathcal{C}_{f_i}\left((\boldsymbol{u}(t), \boldsymbol{y}_0(t))\right) \quad \begin{array}{l} \text{has NO solution for the} \\ \text{unknown variables in it} \end{array} \quad (4.36)$$

directly follows. This part of the proof is based on the falsification of the opposite of the above statement, which is:

$$\exists\, \boldsymbol{u}(t) : \boldsymbol{y}_0(t) \neq \boldsymbol{y}_{f_i}(t) \Rightarrow \mathcal{C}_{f_i}\left((\boldsymbol{u}(t), \boldsymbol{y}_0(t))\right) \quad \begin{array}{l} \text{has at least one solution for the} \\ \text{unknown variables in it } \forall\, \boldsymbol{u}(t). \end{array} \quad (4.37)$$

The inversion of the above statement yields

$$\boldsymbol{y}_0(t) = \boldsymbol{y}_{f_i}(t) \,\forall\, \boldsymbol{u}(t) \Leftarrow \exists\, \boldsymbol{u}(t) : \mathcal{C}_{f_i}\left((\boldsymbol{u}(t), \boldsymbol{y}_0)\right) \quad \begin{array}{l} \text{has NO solution for the} \\ \text{unknown variables in it.} \end{array} \quad (4.38)$$

The above statement is contradictory: Injecting the left part into the right part, one obtains

$$\exists\, \boldsymbol{u}(t) : \mathcal{C}_{f_i}\left(\left(\boldsymbol{u}(t), \boldsymbol{y}_{f_i}(t)\right)\right) \text{ has NO solution for the unknowns in it} \quad (4.39)$$

which is wrong, because it violates the definition of $\boldsymbol{y}_{f_i}(t)$: with $\boldsymbol{y}_{f_i} = \mathcal{C}_{f_i} \circ \boldsymbol{u}(t)$

$$\mathcal{C}_{f_i}\left(\left(\boldsymbol{u}(t), \boldsymbol{y}_{f_i}(t)\right)\right) \quad (4.40)$$

always has a solution. Since the opposite statement was falsified, the original holds true.□

In order to apply Theorem 4.2, the constraint set $\mathcal{C}_{f_i}$ must have the property that specific I/O-pairs do not allow a solution for the unknown variables. This property that specific I/O-pairs do not satisfy - in other words, contradict - the constraint set $\mathcal{C}_{f_i}$ is defined in the following:

Definition 4.3 (Contradictable constraint sets). *A constraint set $\mathcal{C}$ is called contradictable under excitation with the input signal $\boldsymbol{u}(t)$, if a signal $\boldsymbol{u}(t)$ and independently an arbitrary signal $\boldsymbol{y}(t)$ exist, such that the constraint set $\mathcal{C}((\boldsymbol{u}(t), \boldsymbol{y}(t)))$ does not have a solution for the unknown variables $\mathcal{X}$ in it.*

The definition of contradictable constraint sets allows to formulate the following theorem:

Theorem 4.3. *A fault f_i is I/O-detectable if and only if there is an input signal $\boldsymbol{u}(t)$, such that*

- *the constraint set $\mathcal{C}_{f_i}$ is contradictable under excitation with $\boldsymbol{u}(t)$ and*

- *the output signal $\boldsymbol{y}_0(t) = \mathcal{C}_0 \circ \boldsymbol{u}(t)$, which results from the excitation of the fault-free plant with $\boldsymbol{u}(t)$ leads to contradiction with $\mathcal{C}_{f_i} \ \forall \ f_i \neq 0$.*

Proof. For the following proof, Theorem 4.3 is split into two independent assertions, the if-then-part and the only-if-part. Both are proven to hold true separately. Together, they show that Theorem 4.3 holds true.

If-then-part: The following is based on the Theorem 4.2 and Definition 4.3: the I/O-pair that results if the fault-free plant is excited with the input signal $\boldsymbol{u}(t)$ is

$$(\boldsymbol{u}(t), \boldsymbol{y}_0(t)) = (\boldsymbol{u}(t), \mathcal{C}_0 \circ \boldsymbol{u}(t)) . \tag{4.41}$$

If $\mathcal{C}_{f_i}$ is contradictable under excitation with $\boldsymbol{u}(t)$, and the I/O-pair $(\boldsymbol{u}(t), \boldsymbol{y}_0(t))$ contradicts it, by definition $\mathcal{C}_{f_i}((\boldsymbol{u}(t), \boldsymbol{y}_0(t)))$ does not have a solution for the unknown variables $\mathcal{X}$ in it. If there is no solution for the unknown variables $\mathcal{X}$, one can conclude with Theorem 4.2 that the fault f_i is then I/O-detectable.

Only-if-part: According to Definition 4.3, if there is no $\boldsymbol{u}(t)$ under which the constraint set is contradictable, there is no $\boldsymbol{y}(t)$ that leads to contradiction and hence, $\boldsymbol{y}_0(t)$ does not contradict $\mathcal{C}_{f_i}$. With Theorem 4.2 one can conclude that then, the fault is not I/O-detectable and the only-if-part of Theorem 4.3 follows. □

Theorem 4.3 decomposes the necessary and sufficient condition in Theorem 4.2 into two parts: First, the constraint set must have the property that it can be contradicted. While this is a condition on the constraint set $\mathcal{C}_{f_i}$, the second part requires the constraint set $\mathcal{C}_0$ to generate an output signal which leads to this contradiction.

This decomposition is useful because the first part, which is the condition of contradictability, will later be related to structural properties of a constraint set.

Therefore, a main condition in Theorem 4.3 for the I/O-detectability of a fault f_i is the contradictability of $\mathcal{C}_{f_i}$. For a constraint set $\mathcal{C}$ to be contradictable, it is sufficient that one of its subsets is contradictable:

Theorem 4.4 (Contradictability of constraint sets and their subsets). *A constraint set $\mathcal{C}$ is contradictable under excitation with the input signal $\boldsymbol{u}(t)$ if a subset $\mathcal{C}_U \subseteq \mathcal{C}$ is contradictable under excitation with the input signal $\boldsymbol{u}(t)$.*

Proof. If the constraint set $\mathcal{C}_U$ is contradictable under excitation with $\boldsymbol{u}(t)$, the constraint set

$$\mathcal{C} = \mathcal{C}_U \cup \mathcal{C}_{add} \qquad (4.42)$$

which consists of $\mathcal{C}$ and a set $\mathcal{C}_{add}$ of arbitrary additional constraints is contradictable under excitation with $\boldsymbol{u}(t)$. This is because additional constraints can only allow the space of solutions for the unknown variables $\mathcal{X}$ in the constraint set $\mathcal{C}_U$ to become smaller, not larger. From eqn. (4.42) follows that

$$\mathcal{C}_U \subseteq \mathcal{C} \qquad (4.43)$$

holds, and therefore $\mathcal{C}$ is contradictable under excitation with $\boldsymbol{u}(t)$, if $\mathcal{C}_U$ is contradictable under excitation with $\boldsymbol{u}(t)$. $\square$

An important relationship between the I/O-detectability and the contradictability which is based on the insight in Theorem 4.4 is:

Theorem 4.5. *A fault f_i is I/O-detectable, if there is a subset $\mathcal{C}_{U,f_i} \subseteq \mathcal{C}_{f_i}$ which is contradictable under excitation with $\boldsymbol{u}(t)$ with the property that*

$$\mathcal{C}_{U,f_i}\left((\boldsymbol{u}(t), \boldsymbol{y}_0(t))\right), \ \forall \ f_i \neq 0 \qquad (4.44)$$

has no solution for the unknown variables $\mathcal{X}$. The signal $\boldsymbol{y}_0(t)$ is the output of the fault-free plant when excited with the same input signal $\boldsymbol{u}(t)$: $\boldsymbol{y}_0(t) = \mathcal{C}_0 \circ \boldsymbol{u}(t)$.

Proof. From Theorem 4.4 follows that if $\mathcal{C}_{U,f_i} \subseteq \mathcal{C}_{f_i}$ is contradictable under excitation with the input signal $\boldsymbol{u}(t)$, $\mathcal{C}_{f_i}$ is contradictable under excitation with $\boldsymbol{u}(t)$. Therefore, if there is no solution for the unknowns which appear in $\mathcal{C}_{U,f_i}\left((\boldsymbol{u}(t), \boldsymbol{y}_0(t))\right)$ under excitation with $\boldsymbol{u}(t)$, there is also no solution for the unknowns in $\mathcal{C}_{f_i}$ under excitation with $\boldsymbol{u}(t)$. With Theorem 4.3 one can conclude that the fault f_i is then I/O-detectable. $\square$

Theorem 4.5 is an important improvement on Theorem 4.3 for the following reason: Theorem 4.3 requires the entire constraint set $\mathcal{C}_{f_i}$ to be investigated in order to show the I/O-detectability of the fault f_i. Especially for large analytical models this is a difficult task. In contrast to this, Theorem 4.5 allows to infer on the I/O-detectability of the fault f_i only on the basis of a subset $\mathcal{C}_{U,f_i} \subseteq \mathcal{C}_{f_i}$ of the constraints in the analytical model. It is this difference that achieves a significant reduction in the effort of the analysis of the I/O-detectabilty. Futhermore, it is possible to determine such subsets using an abstraction of the analytical model instead of using the constraint set itself. In this manner, Theorem 4.5 paves the way for the global structural analysis in Section 4.5-4.8 and the local structural analysis in Section 5.4-5.5.

Often, it is of interest whether a particular fault is not I/O-detectable. The following theorem provides an answer to this.

Theorem 4.6. *If the variable f_i does not appear in a contradictable subset $\mathcal{C}_{U,f_i} \subseteq \mathcal{C}_{f_i}$, the fault f_i is not I/O-detectable.*

The following proof uses the idea that even although there may be contradictable constraint sets, a fault variable which does not appear in the constraints in these sets may not lead to the contradiction with the sets. The fault is then not I/O-detectable.

Proof. If the fault variable f_i does not appear in an arbitrary contradictable subset $\mathcal{C}_{U,f_i} \subseteq \mathcal{C}_{f_i}$, there is a constraint set $\mathcal{C}_A$ such that

$$\mathcal{C}_0 \supseteq \mathcal{C}_A = \mathcal{C}_{U,f_i} \subseteq \mathcal{C}_{f_i} \tag{4.45}$$

holds. Then,

$$\mathcal{C}_{U,f_i} \circ \boldsymbol{u}(t) = \mathcal{C}_A \circ \boldsymbol{u}(t) \tag{4.46}$$

holds, and the constraint set

$$\mathcal{C}_{U,f_i}((\boldsymbol{u}(t), \mathcal{C}_A \circ \boldsymbol{u}(t))) \tag{4.47}$$

always has at least one solution for the unknown variables $\mathcal{X}$ which appear in it. This is because injecting eqn. (4.46) yields

$$\mathcal{C}_{U,f_i}((\boldsymbol{u}(t), \mathcal{C}_{U,f_i} \circ \boldsymbol{u}(t))) \tag{4.48}$$

which, by definition, always has at least one solution for $\mathcal{X}$. If this holds true for arbitrary contradictable subsets $\mathcal{C}_{U,f_i} \subseteq \mathcal{C}_{f_i}$, Theorem 4.2 allows to conclude that the fault f_i is not I/O-detectable.   □

Although no formal proof is given here, the reverse of Theorem 4.6 also holds: If the fault f_i is I/O-detectable, the variable f_i appears in a contradictable subset $\mathcal{C}_{U,f_i} \subseteq \mathcal{C}_{f_i}$.

For that reason, constraint sets that are contradictable provide necessary conditions to test whether a fault is I/O-detectable. One approach to determine constraint sets which are contradictable independently of the excitation $\boldsymbol{u}(t)$ is the global structural analysis, c.f. Section 4.5 and the following. An approach to identify constraint sets which are contradictable under a specific excitation $\boldsymbol{u}(t)$ is the local structural analysis, c.f. Chapter 5.

The summary of the train of thought in this section is as follows: A fault can be detected on the basis of the input and output signals if its presence has an impact on the behavior of the plant. This is the core of the definition of the I/O-detectability of a fault in Definition 4.2. Lemma 4.1 concludes that this definition is met if there is at least one I/O-pair in the faulty behavior which is not contained in the fault-free behavior. To facilitate further reasoning, this idea was turned around in Lemma 4.2 which states: Definition 4.2 is satisfied if there is at least one I/O-pair in the fault-free behavior which is not contained in the faulty behavior. Lemma 4.1 is used to obtain Theorem 4.1, which states that a fault is I/O-detectble if there is an input signal for which the faulty plant answers with a different output than the fault-free plant. Motivated by Lemma 4.2, Theorem 4.2 gives a necessary and sufficient condition for the I/O-detectabilty of a fault. It consists of the existence of an input signal with the following property: The constraint set which describes the faulty behavior does not have a solution for the unknowns if the I/O-pair which results from the excitation of the fault-free plant with the same input signal is injected into this constraint set. The fundamental property of contradictability, which allows a constraint set not to have a solution for its unknowns is introduced in Definition 4.3. Theorem 4.3 uses this property as a condition on the constraint set describing the faulty system for the I/O-detectability of the fault. Theorem 4.4 states that a constraint set is contradictable if a subset of this constraint set is contradictable. This leads to Theorem 4.5 which states that a fault is I/O-detectable if two conditions are met: First, there needs to be a contradictable subset of the constraint set describing the faulty system. Second, an I/O-pair which results from the excitation of the fault-free system with the same input signal leads to contradiction with this subset. Finally, Theorem 4.2 is used to obtain a condition under which a fault is not I/O-detectable. This condition is given in Theorem 4.6.

In this section, the I/O-detectability of a fault was investigated. Starting from the definition of the I/O-detectability of a fault in Definition 4.2, a number of theorems were formulated, eventually leading to the following finding: The nonexistence of a contradictable subset of the constraint set $\mathcal{C}_{f_i}$ in which the fault variable f_i appears entails that the fault f_i is not I/O-detectable.

4.4   I/O-Discriminability

Whereas the last section discusses the I/O-detectability of a fault, the present section gives an answer to the question whether one can discriminate between the occurrence of two different faults on the basis of the plant's behavior. This is important because it is the main task to be accomplished by the diagnostic unit of an automatic test in service diagnosis.

The problem of distinguishing between the presence of two faults is similar to the problem of detecting the presence of a fault. The latter is solved by distinguishing between the behavior in the fault-free case and the behavior in the faulty case. Similarly, the former can be accomplished by distinguishing between the behavior in the case of fault f_j and the behavior in the case of fault f_i. The solution to both problems is therefore the distinction between two behaviors. However, there is a difference between the two problems: The behavior in the fault-free case is fixed, the behavior in the case of a fault may vary with the magnitude of the fault. Hence, detection is the distinction between a behavior (nominal) and a family of behaviors (faulty). Fault discrimination is the distinction between two families of behaviors (both faulty).

    The property which allows two faults to be distinguished on the basis of the I/O-behavior of the plant is called *I/O-discriminability*. A fault f_i is said to be I/O-discriminable from fault f_j, if it is possible to distinguish the occurrence of fault f_i from the occurrence of fault f_j on the basis of the behavior of the plant. This can be expressed more formally as:

Definition 4.4 (I/O-discriminability). *The fault f_i is called I/O-discriminable from fault f_j, if the behavior of the plant in the case of fault f_i is different from the behavior of the plant in the case of a fault f_j of arbitrary magnitude:*

$$\mathcal{B}_{f_i} \neq \mathcal{B}_{f_j}, \ \forall \ f_i, f_j \neq 0 \tag{4.49}$$

An illustration of the behavior of two I/O-discriminable faults with a given magnitude is depicted in Fig. 4.9.

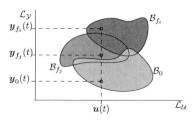

Figure 4.9: The faults f_i and f_j are I/O-discriminable

If there is at least one I/O-pair which is contained in the behavior $\mathcal{B}_{f_i}$ and not contained in the behavior $\mathcal{B}_{f_j}$ for arbitrary magnitudes of f_i and f_j, Definition 4.4 is met. This leads to Lemma 4.3:

Lemma 4.3 (I/O-discriminability). *A fault f_i is I/O-discriminable from fault f_j, if and only if it may entail an I/O-pair which is not contained in the behavior $\mathcal{B}_{f_j}$ which the plant exhibits in the case of a fault f_j of arbitrary magnitude:*

$$\exists \ \left(\boldsymbol{u}, \boldsymbol{y}_{f_i}\right) : \left(\boldsymbol{u}, \boldsymbol{y}_{f_i}\right) \in \mathcal{B}_{f_i} \wedge \left(\boldsymbol{u}, \boldsymbol{y}_{f_i}\right) \notin \mathcal{B}_{f_j}, \ \forall \ f_i, f_j \neq 0 \tag{4.50}$$

Proof. Let $\boldsymbol{u}(t)$ be the signal with which the plant is excited. In the case of fault f_i, the plant answers with the signal $\boldsymbol{y}_{f_i}(t)$, in the case of fault f_j, the plant answers with $\boldsymbol{y}_{f_j}(t)$. The pair $\left(\boldsymbol{u}(t), \boldsymbol{y}_{f_i}(t)\right)$ is contained in $\mathcal{B}_{f_i}$, the pair $\left(\boldsymbol{u}(t), \boldsymbol{y}_{f_j}(t)\right)$ is contained in $\mathcal{B}_{f_j}$. Let $\left(\boldsymbol{u}(t), \boldsymbol{y}_{f_i}(t)\right)$ not be contained in $\mathcal{B}_{f_j}$, independently of the magnitude of f_j. Since two sets are distinct, if there is at least one element, which is contained in one set and not in the other, $\left(\boldsymbol{u}(t), \boldsymbol{y}_{f_i}(t)\right) \notin \mathcal{B}_{f_j} \ \forall \ f_i, f_j \neq 0$ entails $\mathcal{B}_{f_i} \neq \mathcal{B}_{f_j} \ \forall \ f_i, f_j \neq 0$, which is Definition 4.4. $\square$

The close relationship between I/O-detectability and I/O-discriminability is explained in the following remark.

Remark 4.2. *If $f_j = 0$ holds, the problem of discriminating fault f_i from fault f_j is the same as detecting fault f_i. This is because from the qualitative fault models described in Section 3.3.3, the relationship*

$$f_j = 0 \Rightarrow \mathcal{B}_{f_j} = \mathcal{B}_0 \tag{4.51}$$

follows. The distinction between detectability and discriminability is made here to distinguish two problems: namely detecting the presence of a fault and determining its nature. Other authors, c.f. [17], use different fault models which allow them to formulate the distinction between two faults as a fault detection problem on a reduced plant.

The plant's behavior in all fault situations is described by the fault variables and the constraint set in its analytical model. This allows to carry over the reasoning on the discriminability of two faults developed on the basis of the I/O-behaviors $\mathcal{B}_{f_i}$ and $\mathcal{B}_{f_j}$ to the constraint sets $\mathcal{C}_{f_i}$ and $\mathcal{C}_{f_j}$. This is done in the following. Analogous to Theorem 4.1 for the I/O-detectability of a fault f_i, Lemma 4.3 allows to formulate the following theorem for the I/O-discriminability of two faults:

Theorem 4.7. *A fault f_i is I/O-discriminable from fault f_j, if and only if there is an input signal $\boldsymbol{u}(t)$, such that the output signal*

$$\boldsymbol{y}_{f_i}(t) = \mathcal{C}_{f_i} \circ \boldsymbol{u}(t) \tag{4.52}$$

of the plant in the case of the fault f_i is not the same as the output signal

$$\boldsymbol{y}_{f_j}(t) = \mathcal{C}_{f_j} \circ \boldsymbol{u}(t) \tag{4.53}$$

in the case of the fault f_j of arbitrary magnitude: $\boldsymbol{y}_{f_i} \neq \boldsymbol{y}_{f_j} \ \forall \ f_i, f_j \neq 0$.

Proof. By definition

$$\left(\boldsymbol{u}(t), \boldsymbol{y}_{f_i}(t) \right) \in \mathcal{B}_{f_i} \tag{4.54}$$

holds true. With the property of deterministic systems that for each $\boldsymbol{u}(t)$, there is only one $\boldsymbol{y}(t)$,

$$\exists \ \left(\boldsymbol{u}(t), \boldsymbol{y}_{f_i}(t) \right) \notin \mathcal{B}_{f_j}, \ \forall \ f_i, f_j \neq 0 \tag{4.55}$$

must hold, if

$$\exists \ \boldsymbol{u}(t) : \boldsymbol{y}_{f_i}(t) \neq \boldsymbol{y}_{f_j}(t), \ \forall \ f_i, f_j \neq 0 \tag{4.56}$$

holds. Eqn. (4.56) rearranged yields

$$\exists \ \boldsymbol{u}(t) : \mathcal{C}_{f_i} \circ \boldsymbol{u}(t) \neq \mathcal{C}_{f_j} \circ \boldsymbol{u}(t), \ \forall \ f_i, f_j \neq 0. \tag{4.57}$$

If such an input signal $\boldsymbol{u}(t)$ exists, in the case of fault f_i there is an I/O-pair that is not contained in the behavior of the plant in the case of fault f_j, independently of the fault-magnitude f_j. Then, eqn. (4.55) which corresponds to Lemma 4.3, holds. □

Similar to Theorem 4.2 for the I/O-detectability of a fault, a necessary and sufficient condition for the I/O-discriminability of two faults can be given. Like the condition for the I/O-detectability of a fault, the following theorem uses the existence of a solution for the unknown variables in a constraint set:

Theorem 4.8 (Necessary and sufficient condition). *A fault f_i is I/O-discriminable from fault f_j if and only if there is an input signal $\boldsymbol{u}(t)$, such that the constraint set*

$$\mathcal{C}_{f_i}\left(\left(\boldsymbol{u}(t), \mathcal{C}_{f_j} \circ \boldsymbol{u}(t)\right)\right) \tag{4.58}$$

does not have a solution for the unknown variables $\mathcal{X}$ in it, independently of the magnitudes of the faults $f_i, f_j \neq 0$.

Proof. Proving the equivalence of the statement

$$\exists\ \boldsymbol{u}(t) : \boldsymbol{y}_{f_i}(t) \neq \boldsymbol{y}_{f_j}(t) \ \forall\ f_i, f_j \neq 0 \tag{4.59}$$

which is the condition of I/O-detectability in Theorem 4.7 and the statement

$$\exists\ \boldsymbol{u}(t) : \mathcal{C}_{f_i}\left(\left(\boldsymbol{u}(t), \boldsymbol{y}_{f_j}(t)\right)\right) \quad \begin{array}{l}\text{does NOT have a solution for the}\\ \text{unknown variables in it for } f_i, f_j \neq 0 \end{array} \tag{4.60}$$

is done by showing necessity and sufficiency in different steps. The first step consists of proving that from

$$\exists\ \boldsymbol{u}(t) : \mathcal{C}_{f_i}\left(\left(\boldsymbol{u}(t), \boldsymbol{y}_{f_j}(t)\right)\right) \quad \begin{array}{l}\text{does NOT have a solution for the}\\ \text{unknown variables in it for } f_i, f_j \neq 0 \end{array} \tag{4.61}$$

the I/O-discriminability of f_i from f_j follows. By definition,

$$\mathcal{C}_{f_i}\left(\left(\boldsymbol{u}(t), \boldsymbol{y}_{f_i}(t)\right)\right) \tag{4.62}$$

has at least one solution for the unknown variables in it. Therefore, also

$$\boldsymbol{y}_{f_i}(t) = \boldsymbol{y}_{f_j}(t) \ \forall\ \boldsymbol{u}(t) \Rightarrow \mathcal{C}_{f_i}\left(\left(\boldsymbol{u}(t), \boldsymbol{y}_{f_j}(t)\right)\right) \quad \begin{array}{l}\text{has at least one solution for the}\\ \text{unknown variables in it } \forall\ \boldsymbol{u}(t)\end{array} \tag{4.63}$$

holds. Inversion of the above statement yields

$$\exists\ \boldsymbol{u}(t) : \boldsymbol{y}_{f_i}(t) \neq \boldsymbol{y}_{f_j}(t) \Leftarrow \exists\ \boldsymbol{u}(t) : \mathcal{C}_{f_i}\left(\left(\boldsymbol{u}(t), \boldsymbol{y}_{f_j}(t)\right)\right) \quad \begin{array}{l}\text{has NO solution for the}\\ \text{unknown variables in it}\\ \text{for } f_i, f_j \neq 0. \end{array}$$
$$\tag{4.64}$$

The left part is equal to the I/O-discriminability condition in Theorem 4.7 and, therefore, from

$$\exists \, \boldsymbol{u}(t) : \mathcal{C}_{f_i}\left(\left(\boldsymbol{u}(t), \boldsymbol{y}_{f_j}(t)\right)\right) \quad \begin{array}{l} \text{has NO solution for the} \\ \text{unknown variables in it for } f_i, f_j \neq 0, \end{array} \tag{4.65}$$

the I/O-discriminability of the fault f_i from the fault f_j follows. The second step consists of proving that from $\exists \, \boldsymbol{u}(t) : \boldsymbol{y}_{f_j}(t) \neq \boldsymbol{y}_{f_i}(t)$ the statement

$$\exists \, \boldsymbol{u}(t) : \mathcal{C}_{f_i}\left(\left(\boldsymbol{u}(t), \boldsymbol{y}_{f_j}(t)\right)\right) \quad \begin{array}{l} \text{has NO solution for the} \\ \text{unknown variables in it for } f_i, f_j \neq 0 \end{array} \tag{4.66}$$

directly follows. This part of the proof is based on the falsification of the opposite of the above statement, which is:

$$\exists \, \boldsymbol{u}(t) : \boldsymbol{y}_{f_j}(t) \neq \boldsymbol{y}_{f_i}(t) \Rightarrow \mathcal{C}_{f_i}\left(\left(\boldsymbol{u}(t), \boldsymbol{y}_{f_j}(t)\right)\right) \quad \begin{array}{l} \text{has at least one solution for} \\ \text{the unknown variables in it} \\ \forall \, \boldsymbol{u}(t) \text{ for } f_i, f_j \neq 0. \end{array} \tag{4.67}$$

The inversion of this statement yields

$$\boldsymbol{y}_{f_j}(t) = \boldsymbol{y}_{f_i}(t) \; \forall \, \boldsymbol{u}(t) \Leftarrow \exists \, \boldsymbol{u}(t) : \mathcal{C}_{f_i}\left(\left(\boldsymbol{u}(t), y_{f_j}\right)\right) \quad \begin{array}{l} \text{has NO solution for the} \\ \text{unknown variables in it} \\ \text{for } f_i, f_j \neq 0. \end{array} \tag{4.68}$$

The above statement is contradictory: Injecting the left part into the right, one obtains

$$\exists \, \boldsymbol{u}(t) : \mathcal{C}_{f_i}\left(\left(\boldsymbol{u}(t), \boldsymbol{y}_{f_i}(t)\right)\right) \quad \begin{array}{l} \text{has NO solution for the} \\ \text{unknown variables in it for } f_i, f_j \neq 0 \end{array} \tag{4.69}$$

which is wrong, because it violates the definition of $\boldsymbol{y}_{f_i}(t)$. Since the opposite statement was falsified, the original holds true. □

Theorem 4.8 is difficult to handle when analyzing a given analytical model. Therefore, a condition which is based on the constraints in this model is desirable for the analysis of the I/O-discriminability of two faults. This condition is established in the following theorem. It is similar to Theorem 4.3 but makes a statement on the I/O-discriminability of two different faults instead of the I/O-detectability of one fault.

Theorem 4.9. *A fault f_i is I/O-discriminable from fault f_j if there is an input signal $\boldsymbol{u}(t)$, such that*

- *the constraint set $\mathcal{C}_{f_i}$ is contradictable under excitation with $\boldsymbol{u}(t)$ and*

- *the output signal $\boldsymbol{y}_{f_j}(t) = \mathcal{C}_{f_j} \circ \boldsymbol{u}(t)$, which results from the excitation of the plant in the case of fault f_j with $\boldsymbol{u}(t)$ leads to contradiction with $\mathcal{C}_{f_i}$ for all $f_i, f_j \neq 0$.*

Proof. The following proof is based on Definition 4.3 and Theorem 4.8. The I/O-pair that results if the plant with fault f_j is excited with the input signal $\boldsymbol{u}(t)$ is

$$\left(\boldsymbol{u}(t), \boldsymbol{y}_{f_j}(t)\right) = \left(\boldsymbol{u}(t), \mathcal{C}_{f_j} \circ \boldsymbol{u}(t)\right). \tag{4.70}$$

If $\mathcal{C}_{f_i}$ is contradictable under excitation with $\boldsymbol{u}(t)$, and the I/O-pair $\left(\boldsymbol{u}(t), \boldsymbol{y}_{f_j}(t)\right)$ contradicts it, $\mathcal{C}_{f_i}\left(\left(\boldsymbol{u}(t), \boldsymbol{y}_{f_j}(t)\right)\right)$ does not have a solution for the unknown variables $\mathcal{X}$ in it. If there is no solution for the unknown variables $\mathcal{X}$, with Theorem 4.8 one can conclude that the fault f_i is I/O-discriminable from fault f_j. $\qquad\square$

An important observation in Theorem 4.4 which led to Theorem 4.5 for the I/O-detectability of a fault was the insight that a constraint set is contradictable under excitation with $\boldsymbol{u}(t)$ if one of its subsets is contradictable under excitation with $\boldsymbol{u}(t)$. This result allows to refine the condition for the I/O-discriminability of two faults:

Theorem 4.10. *A fault f_i is I/O-discriminable from fault f_j, if there is a subset $\mathcal{C}_{\mathrm{U},f_i} \subseteq \mathcal{C}_{f_i}$ with the property that*

$$\mathcal{C}_{\mathrm{U},f_i}\left(\left(\boldsymbol{u}(t), \boldsymbol{y}_{f_j}(t)\right)\right) \tag{4.71}$$

has no solution for the unknown variables $\mathcal{X}$ under excitation with the input signal $\boldsymbol{u}(t)$ for arbitrary magnitudes of the faults $f_i, f_j \neq 0$.

Proof. From Theorem 4.4 follows that if $\mathcal{C}_{\mathrm{U},f_i} \subseteq \mathcal{C}_{f_i}$ is contradictable under excitation with the input signal $\boldsymbol{u}(t)$, the constraint set $\mathcal{C}_{f_i}$ is contradictable under excitation with $\boldsymbol{u}(t)$. Therefore, the first part of Theorem 4.9 is satisfied. If there is no solution for the unknowns which appear in $\mathcal{C}_{\mathrm{U},f_i}\left(\left(\boldsymbol{u}(t), \boldsymbol{y}_{f_j}(t)\right)\right)$ under excitation with $\boldsymbol{u}(t)$ for arbitrary magnitudes of f_j, there is no solution for the unknowns in $\mathcal{C}_{f_i}\left(\left(\boldsymbol{u}(t), \boldsymbol{y}_{f_j}(t)\right)\right)$ under excitation with $\boldsymbol{u}(t)$ for arbitrary magnitudes of f_j. Then, $\mathcal{C}_{f_i}$ is contradicted and the second part of Theorem 4.9 is satisfied. Then, the faults f_i and f_j are I/O-discriminable.$\square$

Similar to the improvement of Theorem 4.5 on Theorem 4.3 for the I/O-detectability, Theorem 4.10 is an important improvement on Theorem 4.9 for the I/O-discriminability. This is the case for the following reason: In order to investigate the I/O-discriminability of two faults with the help of Theorem 4.9, the entire analytical model needs to be investigated. This is usually a difficult task for large analytical models. Instead, with Theorem 4.10, one can infer on the I/O-discriminability of the faults f_i and f_j on the basis of a constraint subset $\mathcal{C}_{\mathrm{U},f_i} \subseteq \mathcal{C}_{f_i}$. Such a subset may be found using structural analysis. These approaches are detailed in Section 4.5-4.8 and Section 5.4-5.5.

The reasoning for the conditions of the I/O-discriminability of two faults is similar to the reasoning in the previous section which focused on conditions for the I/O-detectability. The train of thought in the present section can be summarized as follows: The definition of the I/O-discriminability of two faults in Definition 4.4 is based on the following idea: If two faults are to be distinguished on the basis of the plant behavior, this behavior must be different in the case of the two faults. Lemma 4.3 states that this definition is met for two faults, f_i and f_j, if there is at least one I/O-pair which is contained in the behavior $\mathcal{B}_{f_i}$ but which is not contained in $\mathcal{B}_{f_j}$ for arbitrary fault-magnitudes. This intermediate result is used to formulate Theorem 4.7, in which the following condition for the I/O-discriminability of two faults is given: Two faults are I/O-discriminable if there is an input signal, which entails different output signals in the case of the two faults for arbitrary fault-magnitudes. A necessary and sufficient condition for this is given in Theorem 4.8: Let the I/O-pair which results from the plant subject to fault f_j be injected into the constraint set $\mathcal{C}_{f_i}$ which describes the behavior in the case of fault f_i. If there is an input signal such that this constraint set does not have a solution for the unknowns in it for arbitraty fault-magnitudes of f_i and f_j, the faults are I/O-discriminable. A constraint set may only have no solution for the unknown variables in it if it is contradictable, c.f. Definition 4.3. This allows to reformulate the condition for the I/O-discriminability which is done in Theorem 4.9: The two faults f_i and f_j are I/O-discriminable, if the constraint set $\mathcal{C}_{f_i}$ is contradictable and the I/O-pair which results from the plant in the case of fault f_j contradicts $\mathcal{C}_{f_i}$. Theorem 4.9 and the result in Theorem 4.4 eventually lead to Theorem 4.10 which states that the two faults are I/O-discriminable, if there is a contradictable subset of $\mathcal{C}_{f_i}$ and the I/O-pair which results from the plant in the case of fault f_j leads to contradiction of this subset.

This section discusses the I/O-discriminability, which is the property that allows to distinguish faults on the basis of the I/O-behavior of the plant. The theorems 4.7-4.10 develop the train of thought from the definition of the I/O-discriminability to conditions of the existence of specific subsets of the constraint set in the analytical plant model. These conditions concern contradictable constraint sets. The subsets allow to determine whether two faults are I/O-discriminable. Section 4.8 and Chapter 5 discuss techniques to provide such constraint sets.

4.5 Global Structural Model of Dynamical Systems

In this section, the global structure graph is introduced. It will serve as a tool to determine constraint sets which are typically contradictable under an *arbitrary* excitation. The main concepts are taken from [22] and [76] and adapted to the model used in this thesis.

4.5.1 Global Structure Graph

A structure graph of a plant is a graph which qualitatively represents the couplings among the variables via the constraints in the analytical model of a plant. It is therefore an abstraction of the analytical model of a plant. If this description holds in an arbitrary operating region, it is referred to as the *global structure graph* in this thesis.

Structural modeling of fault-free systems can be found in a large number of publications, for example [22] and the references therein. In contrast to these references, the approach chosen in this thesis, is to consider the structural model of dynamical systems subject to faults. Similar to [52], [53], [77] and [80], the faults are represented by variable-vertices in the structure graph. In the following, this representation is briefly reviewed. The global structure graph

$$G = (\mathcal{C} \cup \mathcal{Z}, \mathcal{E}) \quad \text{with} \quad \mathcal{Z} = \mathcal{X} \cup \mathcal{K} \cup \mathcal{F} \qquad (4.72)$$

is a bipartite graph with two kinds of vertices, which represent the variables $\mathcal{Z}$ and the constraints $\mathcal{C}$ of the analytical model. There is an edge $e \in \mathcal{E}$ between a variable-vertex $z_j \in \mathcal{Z}$ and a constraint-vertex $c_i \in \mathcal{C}$, if the variable z_j appears in that constraint c_i. In order to express that the above rule was applied to determine the global structure graph of an analytical model in which the constraint set $\mathcal{C}$ describes the behavior, $G(\mathcal{C})$ is written. In figures which depict structure graphs, bars are used for the vertices which represent constraints and circles are used for the variable-vertices. In that way, the bipartite nature of the graph can easily be perceived.

A representation of the global structure graph is its *incidence matrix* $\boldsymbol{M}$. In this matrix, the rows $\boldsymbol{m}^T$ correspond to the constraint-vertices and the columns to the variable-vertices. The ij-th element m_{ij} of the global incidence matrix $\boldsymbol{M}$ is defined as

$$m_{ij} = \begin{cases} 1, & \text{if } z_j \text{ appears in } c_i, \\ 0, & \text{otherwise.} \end{cases} \qquad (4.73)$$

Hence, $m_{ij} = 1$ corresponds to an edge between the vertices c_i and z_j in the global structure graph. In analogy to the partition eqn. (3.2) and eqn. (3.3) of the variables, the incidence matrix can be written as

$$\boldsymbol{M} = [\boldsymbol{M}_{\mathcal{X}} \ \boldsymbol{M}_{\mathcal{K}} \ \boldsymbol{M}_{\mathcal{F}}]. \qquad (4.74)$$

This results in a matrix whose structure is depicted in Fig. 4.10. The matrices describe on which part of the plant – that is on which constraints in the analytical model – the inputs and the outputs interact with the plant $(\boldsymbol{M}_{\mathcal{K}})$ and on which part of the plant the faults may have an impact $(\boldsymbol{M}_{\mathcal{F}})$.

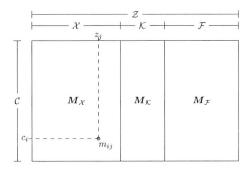

Figure 4.10: Incidence matrix

Example 4.1 (Global structure graph of Plant A). *The global structure graph of Plant A which was introduced in Section 3.4 is depicted in Fig. 4.11.*

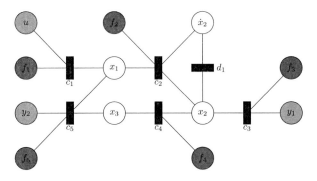

Figure 4.11: Global structure graph G of Plant A

Example 4.2 (Incidence matrix of the global structure graph of Plant A).
The incidence matrix of the global structure graph of Plant A which is depicted in Example 4.1 is given below. For ease of readability, zeros are omitted.

Table 4.1: Incidence matrix of Plant A

	x_1	x_2	$\dot{x}_2$	x_3	u	y_1	y_2	f_1	f_2	f_3	f_4	f_5
m_1^T	1				1			1				
m_2^T	1	1	1						1			
m_3^T		1				1				1		
m_4^T		1		1							1	
m_5^T	1			1			1					1
$m_{d_1}^T$		1	1									

$$\underbrace{\hspace{3.5cm}}_{M_\chi} \quad \underbrace{\hspace{2.5cm}}_{M_\kappa} \quad \underbrace{\hspace{4cm}}_{M_\mathcal{F}}$$

Variables in constraint sets.   The graph G can be used to determine how the values of the variables propagate through the plant. For this purpose, the $\mathrm{var}^G(\cdot)$-operator is defined in the following. It allows to determine which variables interfere with a set of constraints $\mathcal{C}$ according to the structure graph G:

Definition 4.5 $\big(\mathrm{var}^G(\cdot)$**-operator**$\big)$.   *The operator* $\mathrm{var}^G(\cdot)$ *determines the set of all variables $\mathcal{Z}$ that interfere with the set of constraints $\mathcal{C}$ according to the structure graph G. It therefore matches the powerset of constraints to the powerset of variables:*

$$\mathrm{var}^G(\mathcal{C}) : \ 2^{\mathcal{C}} \mapsto 2^{\mathcal{Z}}. \tag{4.75}$$

The $\mathrm{var}^G(\mathcal{C})$-operator can easily be implemented on an incidence matrix $\boldsymbol{M}$ by the following procedure:

1. Determine the rows $\boldsymbol{m}_i^T$ in the incidence matrix $\boldsymbol{M}$ of the graph G which correspond to the constraints in the set $c_i \in \mathcal{C}$.

2. Determine the columns which have nonzero entries in the rows found in Step 1.

3. Determine the variables which correspond to the columns found in Step 2. These are the variables in $\mathrm{var}^G(\mathcal{C})$.

Example 4.3 $\big(\mathrm{var}^G(\cdot)$**-operator applied to a constraint set from Plant A**$\big)$.   *Let the constraint set $\mathcal{C}_{2,3} = \{c_2, c_3\}$ be taken from the analytical model of Plant A. Then, with the global incidence matrix $\boldsymbol{M}$ from Example 4.2 which represents the global structure graph G of Plant A, one finds*

$$\mathrm{var}^G(\mathcal{C}_{2,3}) = \{x_1, x_2, \dot{x}_2, y_1, f_2, f_3\}. \tag{4.76}$$

Matching.   Selecting the edge between the constraint-vertex c_i and the variable-vertex z_j in a structure graph G can be used to express that a rearrangement of the constraint

c_i is used to compute the value of the variable z_j. This rearrangement can be obtained by solving g_i for z_j:

$$z_j = g_i^{-1}(z_k, ...). \tag{4.77}$$

The other variables

$$\mathcal{Z}_{\text{Rest}} = \text{var}^G(c_i) \setminus z_j \tag{4.78}$$

which appear in g_i^{-1} are necessary to determine the value of z_j. A variable z_j is called *matched* if the edge between it and c_i in the graph is selected. In order to compute z_j, it may be necessary to determine the value of the unknown variables in the set $\mathcal{Z}_{\text{Rest}}$. This in turn makes it necessary that these variables are matched and, hence, other edges in the graph are selected. In this manner, a way to compute the value of the variable z_j from known variables can be obtained.

Further edges that are selected must have the property that they are not adjacent to the vertex representing the constraint c_i and the vertex representing the variable z_j.

The reasons for that are the following:

a) A rearrangement of c_i can only be used to determine the value of *one* variable. No other variable's value can be determined from c_i and, therefore, no edge that connects c_i and any other variable than z_j may be selected.

b) If the value of z_j is already determined by a rearrangement of c_i, it is not necessary to determine it by another constraint. Therefore no edge connecting z_j to another constraint may be selected.

If edges are selected in the global structure graph in order to determine the value of z_j and the two properties a) and b) are respected, one obtains a set of edges which have the property that no two edges share a vertex. This motivates the definition of a specific property of a set of edges in the global structure graph:

Definition 4.6 (Matching). *A subset* $\mathcal{M} \subset \mathcal{E}$ *of the edges in a structure graph G is called* matching, *if no two edges in $\mathcal{M}$ have a common vertex.*

With this definition, a matching corresponds to a set of selected nonzero entries in the incidence matrix, with the property that there is only a single "1" selected per row and per column.

The edge e which matches a variable to a constraint means that the value of the variable is determined from the constraint. If this requires the values of other variables to be computed, these variables need to be matched as well. An entire matching $\mathcal{M}$ therefore defines an order of computation for a variable. This order is sometimes called alternating chain, c.f. [22].

Since a matching can be understood as the order in which variables are computed in a constraint set, it describes for which variable a constraint needs to be solved to compute the unknown variables. This can be visualized by directed edges in the structure graph. A matching therefore assigns directions to some of the edges in the graph. Obviously, solving a constraint for a specific variable uniquely is not always possible. This is taken into account by the concept of directed structure graphs which is described in Section 4.6. In the following, an example of a matching is given.

If all unknown variables need to be determined, edges between all vertices corresponding to an unknown variable and a constraint need to be selected. Also, if all constraints shall be used, edges adjacent to all constraint-vertices need to be selected. Using the operator $|\cdot|$ for the cardinality of a set, this leads to the following definition:

Definition 4.7 (Complete and maximal matching). *A matching is called* complete *w.r.t. $\mathcal{C}$ if $|\mathcal{M}| = |\mathcal{C}|$ holds. It is called complete w.r.t. $\mathcal{Z}$ if $|\mathcal{M}| = |\mathcal{Z}|$ holds. A matching is called* maximal, *if adding an arbitrary edge to it destroys the no-common-vertex-property.*

Example 4.4 (Matching on the graph of Plant A). *A complete matching on the global structure graph of Plant A is described with the help of the global incidence matrix below. Selected edges are indicated with* ①*, zeros are omitted for ease of readibility.*

Table 4.2: Matching on the global structure graph G of Plant A

	x_1	x_2	$\dot{x}_2$	x_3	u	y_1	y_2	f_1	f_2	f_3	f_4	f_5
m_1^T	①				1			1				
m_2^T	1	①	1						1			
m_3^T		1				1				1		
m_4^T		1		①							1	
m_5^T	1			1			1					1
$m_{d_1}^T$		1	①									

$$\underbrace{\qquad\qquad}_{M_\chi} \quad \underbrace{\qquad}_{M_\kappa} \quad \underbrace{\qquad\qquad}_{M_\mathcal{F}}$$

4.5.2 Canonical Decomposition of the Structure Graph

The Dulmage-Mendelsohn-Decomposition [46] allows the canonical decomposition of a bipartite graph. If the two vertex sets correspond to constraint-vertices and variable-vertices respectively, the Dulmage-Mendelsohn-Decomposition of the graph G results in the partition of the constraint set $\mathcal{C}$ into the sets $\mathcal{C}^+$, $\mathcal{C}^0$ and $\mathcal{C}^-$ and in the partition of the unknown variables into the variable sets $\mathcal{X}^+$, $\mathcal{X}^0$ and $\mathcal{X}^-$. This is done by row and column permutation of the incidence matrix describing the graph G. A sketch of the result obtained with the decomposition is depicted in Fig. 4.12.

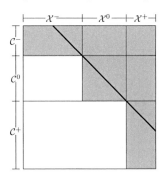

Figure 4.12: Canonical Dulmage-Mendelsohn-Decomposition of M_χ

In this figure, white areas correspond to zeros in the incidence matrix, gray areas correspond to ones or zeros in the incidence matrix and the thick line corresponds to a matching [22].

The partitions of the constraints and unknown variables have the property that on the global structure graph of

- $\mathcal{C}^+, \mathcal{X}^+$, there is a complete matching w.r.t. the variables $\mathcal{X}^+$ but not w.r.t. the constraints $\mathcal{C}^+$,

- $\mathcal{C}^0, \mathcal{X}^0$, there is a complete matching w.r.t. the variables $\mathcal{X}^0$ and w.r.t. the constraints $\mathcal{C}^0$,

- $\mathcal{C}^-, \mathcal{X}^-$, there is a complete matching w.r.t. the constraints $\mathcal{C}^-$ but not w.r.t. the variables $\mathcal{X}^-$.

Since the decomposition of a constraint set into these three sets is an important tool for the structural analysis, an operator for the decomposition is defined:

Definition 4.8 (Canonical decomposition). *The operator* $\mathrm{DM}(\mathcal{C})$ *realizes a partition of the constraint set* $\mathcal{C}$ *into the three subsets*

$$\{\mathcal{C}^+, \mathcal{C}^0, \mathcal{C}^-\} = \mathrm{DM}(\mathcal{C}) \tag{4.79}$$

with the property that

- *there is a complete matching w.r.t. the variables* $\mathcal{X}^+$ *but not w.r.t. the constraints* $\mathcal{C}^+$

- *there is a complete matching w.r.t. the variables* $\mathcal{X}^0$ *and w.r.t. the constraints* $\mathcal{C}^0$

- *there is a complete matching w.r.t. the constraints* $\mathcal{C}^-$ *but not w.r.t. the variables* $\mathcal{X}^-$

on the structure graph G.

The constraint set $\mathcal{C}^+$ is also called the structurally overconstrained part of the constraint set $\mathcal{C}$, $\mathcal{C}^0$ is called the justconstrained part and $\mathcal{C}^-$ is called the underconstrained part. In the following, the superscripts $+, 0, -$ will be used to indicate that the operator $\mathrm{DM}(\mathcal{C})$ was used on the constraint set $\mathcal{C}$ to determine the over-, just-, or underconstrained part of the constraint set respectively.

4.5.3   Global Structural Properties of Constraint Sets

In this section, the global structural properties of constraint sets are introduced. These properties play an important role in the analysis of the global structure graph, the result of which is linked to the I/O-diagnosability properties later.

The following definition is in accordance with [76]:

Definition 4.9 (Global structural properties of constraint sets). *A set of constraints $\mathcal{C}$ is called*

Globally Structurally Justconstrained (GSJ), *if there is a matching that is complete w.r.t. the unknown variables and w.r.t. the constraints on the global structure graph G of the constraint set $\mathcal{C}$. Such a constraint set is denoted by $\mathcal{C}_{\mathrm{GSJ}}$.*

Globally Structurally Overconstrained (GSO), *if the global structure graph G of the constraint set $\mathcal{C}$ indicates that there are more constraints in $\mathcal{C}$ than unknowns:*

$$|\mathcal{C}| > |\mathcal{X} \cap \mathrm{var}^G(\mathcal{C})| \qquad (4.80)$$

Such a constraint set is denoted by $\mathcal{C}_{\mathrm{GSO}}$.

Globally Proper Structurally Overconstrained (GPSO), *if it is globally structurally overconstrained, and*

$$\mathcal{C} = \mathcal{C}^+ \qquad (4.81)$$

holds. Such a constraint set is denoted by $\mathcal{C}_{\mathrm{GPSO}}$.

Globally Minimal Structurally Overconstrained (GMSO), *if it is globally proper structurally overconstrained, but none of its proper subsets is a globally proper structurally overconstrained set. It is denoted by $\mathcal{C}_{\mathrm{GMSO}}$.*

Approaches for finding GMSOs can be found in a number of publications, for example in [102] and [103]. The basic algorithm in [76] makes use of the concept of the *structural redundancy* $\bar{\varphi}$. The scalar $\bar{\varphi}$ is the difference between the number of constraints in the overconstrained part $\mathcal{C}$ of the constraint set and the number of unknown variables which appear in the constraints in the set $\mathcal{C}^+$. If $\bar{\varphi}(\mathcal{C}) = 1$ holds, the constraint set is GMSO. If the variables which appear in the constraints in a constraint set $\mathcal{C}$ are determined with the operator $\mathrm{var}^G(\mathcal{C})$, the global structure graph G implies a structural redundancy of the constraint set $\mathcal{C}$. This is because the operator $\mathrm{var}^G(\mathcal{C})$ makes use of the graph G. Similar to [76], we therefore define the structural redundancy implied by the structure graph G as

$$\bar{\varphi}(\mathcal{C}) = |\mathcal{C}^+| - (\mathcal{X} \cap \mathrm{var}^G(\mathcal{C}^+)). \qquad (4.82)$$

The following algorithm to compute GMSOs is taken from [76] and adapted to the notation used in this thesis. Its input is therefore not only a constraint set but its global structure graph as well.

The idea behind this algorithm is to successively remove constraints from a constraint set until removing a further constraint results in the structural redundancy implied by $G(\mathcal{C})$ being one. Note that more efficient algorithms are presented in [76]. However, since the focus of this thesis is not numerical efficiency and the simplicity of the below algorithm facilitates the comprehension, the more efficient approaches in [76] are not pursued here.

Algorithm 1 [FindMSO($\mathcal{C}, G$)]:

Input: A set of constraints $\mathcal{C}$ and the global structure graph $G(\mathcal{C})$.

Step 1:   if $\bar{\varphi}(\mathcal{C}) = 1$ then
$\qquad\qquad\mathcal{A}_{\mathrm{GMSO}} := \mathcal{C}$
$\qquad$ else

Step 2: $\mathcal{A}_{\mathrm{GMSO}} := \emptyset$;
$\qquad\qquad$ for each constraint $c_i \in \mathcal{C}$ do
$\qquad\qquad\qquad \mathcal{C}' := (\mathcal{C} \setminus \{c_i\})^+$;
$\qquad\qquad\qquad \mathcal{A}_{\mathrm{GMSO}} := \mathcal{A}_{\mathrm{GMSO}} \cup \mathrm{FINDMSO}(\mathcal{C}', G(\mathcal{C}'))$;
$\qquad\qquad$ end for
$\qquad$ end if
$\qquad$ return $\mathcal{A}_{\mathrm{GMSO}}$

Result: A family of sets $\mathcal{A}_{\mathrm{GMSO}}$ the elements of which are the GMSOs $\mathcal{C}_{\mathrm{GMSO}}$

Note that in order to determine the constraint set $\mathcal{C}'$ and the global structure graph $G(\mathcal{C}')$ of this constraint set, it is not necessary to determine a new structure graph by analyzing which variables appear in the constraints in the set $\mathcal{C}'$. The constraint set $\mathcal{C}' = (\mathcal{C} \setminus \{c_i\})^+$ can be obtained with the help of the incidence matrix of the global structure graph of $\mathcal{C} \setminus \{c_i\}$. The incidence matrix of the graph $G(\mathcal{C} \setminus \{c_i\})$ results if the constraint-vertex c_i and all edges adjacent to c_i are removed from the global structure graph of $\mathcal{C}$. This corresponds to eliminating the row that corresponds to c_i from the incidence matrix. The globally proper structurally overconstrained part $(\mathcal{C} \setminus \{c_i\})^+$ can then be determined with the Dulmage-Mendelsohn-Decomposition $\mathrm{DM}(\mathcal{C} \setminus \{c_i\})$.

Example 4.5 (GMSOs of Plant A).   *Analyzing the constraint set $\mathcal{C}$ of Plant A and its global structure graph G with the algorithm* FINDMSO *yields a total of four GMSOs:*

$$\mathcal{A}_{\mathrm{GMSO}} \begin{cases} \mathcal{C}_{\mathrm{GMSO},1} & = \{c_1, c_2, c_3, d_1\} \\ \mathcal{C}_{\mathrm{GMSO},2} & = \{c_1, c_2, c_4, c_5, d_1\} \\ \mathcal{C}_{\mathrm{GMSO},3} & = \{c_1, c_3, c_4, c_5\} \\ \mathcal{C}_{\mathrm{GMSO},4} & = \{c_2, c_3, c_4, c_5, d_1\} \end{cases}$$

If an arbitrary constraint is removed from a GMSO, the property that there is no complete matching w.r.t. the constraints is lost. The resulting constraint set is globally structurally justconstrained.

Remark 4.3.   *For deterministic systems, the hierarchical order*

$$\mathcal{C} \supseteq \mathcal{C}_{\mathrm{GSO}} \supseteq \mathcal{C}_{\mathrm{GPSO}} \supseteq \mathcal{C}_{\mathrm{GMSO}} \supset \mathcal{C}_{\mathrm{GSJC}} \qquad (4.83)$$

between sets of constraints generally holds.

In this section, the global structure graph was introduced. Properties of constraint sets which can be verified using this graph were given. An algorithm to determine a specific kind of constraint sets, called GMSO, based on the global structure graph was presented. Such constraint sets form the basis for the determination of global residual generators in Section 4.7.

4.6   Directed Global Structure Graph

In this section, a specific extension of the structure graph is introduced which takes into account that a constraint may not have a unique solution for a specific variable occurring in the constraint. Similar to the concepts of invertibility in [41] and [42] and the concept of calculability in [65], the information, for which variable a constraint can be solved is introduced in the global structure graph. This extension consists of forbidding some edges to be used in a matching. The directed global structure graph allows to formulate a method to determine global residual generators in Section 4.7.

4.6.1   Uniqueness of Constraints Solved for a Variable (Causality)

An edge between a constraint-vertex c_i and a variable-vertex z_j in a matching is associated with solving the constraint c_i for the variable z_j in order to determine the value of z_j from the values of the other variables. This is, in some cases, not uniquely possible. These cases are in particular:

- Differential constraints: Without the knowledge of the initial value it is not possible to determine $z(t)$ from its derivative $\dot{z}(t)$. The constraint

$$c_i : \quad 0 = \dot{z}(t) - \frac{\mathrm{d}}{\mathrm{d}t} z(t) \qquad (4.84)$$

  cannot be solved uniquely for $z(t)$. Contrariwise, it is possible to determine the derivative $\dot{z}(t)$ from the variable $z(t)$.

- Algebraic constraints with more than one solution: If a constraint is solved for a variable, different solutions may be possible. An illustrative example of this situation is the quadratic constraint

$$c_i : \quad 0 = (z(t))^2 - x(t). \qquad (4.85)$$

  Since there are two solutions for $z(t)$, it is not possible, to compute $z(t)$ uniquely from $x(t)$ by a rearrangement of c_i. Contrariwise, it is possible to determine $x(t)$ uniquely from $z(t)$ on the basis of $z(t)$.

- Inequalities: Obviously, an inequality does not allow to determine the value of a variable by rearranging the inequality. The example

$$c_i : \quad z(t) < 0 \qquad (4.86)$$

  illustrates this fact.

The above cases show that not all constraints may be used to compute the value of all the variables occurring in them. Nevertheless, if the value of a variable is known, it may be injected into every constraint in which the variable appears.

This motivates the idea to include the information whether a matched constraint provides a unique solution for the matched variable in the structure graph itself.

4.6.2   Directed Global Structure Graph

Rearranging a constraint in order to determine the value of one variable from the other variables interfering in the constraint corresponds to matching the variable to the constraint. This is sometimes illustrated by a directed edge in the global structure graph, c.f. [22]. Following the reasoning that some variables may not be computed uniquely from a constraint, an edge between such a constraint-variable pair may not be part of a matching.

An edge, which connects the constraint-vertex c_i and the variable-vertex z_j in a global structure graph is called *acausal* in the following, if no rearrangement of c_i is unique w.r.t. z_j.

Although acausal edges may not be part of a matching, they cannot be left out in a structural representation of the constraint set. This is because a variable interferes with every constraint which is adjacent to the variable in the structure graph. Instead, the acausal edge can be represented by a directed edge. Note that this is done prior to the assignment of directions by a matching. In Fig. 4.13 both cases are depicted in order to illustrate the difference between the direction of an acausal edge and the direction of an edge assigned by a matching.

The direction assigned by a matching describes which variable is computed from which constraint. Therefore, the direction of the edge is always from a constraint-vertex towards a variable-vertex in that case. In contrast, causality describes which variable cannot be determined uniquely from a constraint. The value of the variable may only be used in a constraint. Therefore, the direction of an edge which expresses causality is from a variable-vertex towards a constraint-vertex.

Figure 4.13: Difference between the direction of edges assigned by a matching (left) and due to causality (right)

If the investigation whether an edge is causal, is conducted for all edges in a global structure graph, one finds the directed global structure graph according to the following definition.

Definition 4.10 (Directed global structure graph). *The global structure graph of a constraint set $\mathcal{C}$, in which all acausal edges are represented by directed edges, is called* directed global structure graph.

In Example 4.6, the directed global structure graph of Plant A is depicted. In the following, G is used for the directed global structure graph. Directed edges are represented by a "-1" instead of a "1" in the incidence matrix of the graph. Other authors [22] use an $"X"$ in the incidence matrix to mark edges with a similar meaning.

Example 4.6 (Directed global structure graph of Plant A). *Figure 4.14 shows the directed global structure graph of Plant A. The difference with respect to the global structure graph is the directed edge from the unknown variable x_2 to the differential constraint d_1.*

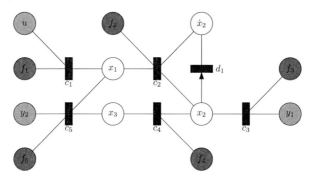

Figure 4.14: Directed global structure graph of Plant A

In Tab. 4.3 the incidence matrix corresponding to the directed global structure graph of Plant A is given. The difference with respect to Tab. 4.1 is the "-1" in row $\boldsymbol{m}_{d_1}^T$.

Table 4.3: Incidence matrix of the directed global structure graph of Plant A

	x_1	x_2	$\dot{x}_2$	x_3	u	y_1	y_2	f_1	f_2	f_3	f_4	f_5
$\boldsymbol{m}_1^T$	1				1			1				
$\boldsymbol{m}_2^T$	1	1	1						1			
$\boldsymbol{m}_3^T$		1				1				1		
$\boldsymbol{m}_4^T$		1		1							1	
$\boldsymbol{m}_5^T$	1			1			1					1
$\boldsymbol{m}_{d_1}^T$		-1	1									

$$\underbrace{}_{\boldsymbol{M}_{\mathcal{X}}} \qquad \underbrace{}_{\boldsymbol{M}_{\mathcal{K}}} \qquad \underbrace{}_{\boldsymbol{M}_{\mathcal{F}}}$$

The acausal edges describe that the solution of a constraint for a variable is not unique. Any order of computation which contains solutions that are not unique is generally not unique itself. Because a matching defines an order of computation, this computation can only be unique if it does not contain acausal edges. This leads to the following definition which will play an important role in the design of residual generators.

Definition 4.11 (Causal matching). *A matching $\mathcal{M}$ is called* causal matching, *if it does not contain any acausal edges.*

In this section, the concept of the directed global structure graph was introduced. This graph is an extension of the global structure graph which incorporates the information whether a variable may be computed uniquely from a constraint. This is an important property which is used in the following section for the design of global residual generators.

4.7   Determining Global Residual Generators

In this section, a method to determine global residual generators is presented. It makes use of the algorithm FINDMSO and the directed global structure graph.

A residual generator aims at showing the inconsistency of an observed I/O-pair with the nominal behavior of the plant. From Theorem 4.5 it is known that one condition for a fault to be I/O-detectable is the existence of a constraint set $\mathcal{C}_{\mathrm{U},f_i} \subseteq \mathcal{C}_{f_i}$ which is contradictable and in which there are constraints in which the fault variable f_i appears. This information can be obtained with the global structure graph as well: $f_i \in \mathrm{var}^G(\mathcal{C}_{\mathrm{U},f_i})$ must hold.

Two observations motivate the idea to use GMSOs, in which the variable f_i appears as the sets $\mathcal{C}_{\mathrm{U},f_i}$ for the construction of global residual generators:

First, a GMSO is typically contradictable, independently of the excitation[2] $\boldsymbol{u}(t)$. This is because a GMSO is typically a constraint set for which independently of the nature of $\boldsymbol{u}(t)$, there is an arbitrary signal $\boldsymbol{y}(t)$ so that there is no solution for the unknown variables in the GMSO.

Second, only a variable that appears in a constraint set may have an impact on the existence of a solution for the unknowns appearing in this constraint set. Inconsistency of an observed I/O-pair with a fault-free constraint set can only be caused by a fault variable which appears in this constraint set, c.f. Theorem 4.6. This is because determining whether $(\boldsymbol{u}(t), \mathcal{C}_0 \circ \boldsymbol{u}(t))$ is contained in $\mathcal{B}_{f_i}$ (which was used in the proofs for Lemma 4.2) is the same as determining whether $(\boldsymbol{u}(t), \mathcal{C}_{f_i} \circ \boldsymbol{u}(t))$ is contained in $\mathcal{B}_0$. Since it is advantageous for fault discrimination if a residual is sensitive to only few faults, contradictable constraint sets in which only few fault variables appear should be used for the design of residual generators. GMSOs with a causal matching are the constraint sets which have the smallest cardinality and which are typically contradictable under excitation with an

---

[2]As a matter of fact, another property is necessary for this, which is explained later on: there needs to be a causal matching.

arbitrary input signal at the same time. The small cardinality entails a small number of fault variables which appear in the GMSOs. Therefore, global residuals with good discrimination properties can be presumably obtained from GMSOs. A more detailed analysis of the relationship between structural properties of the analytical model of a plant and the I/O-properties evaluated by a residual generator is conducted in Section 4.8.

The basic idea to determine a global residual generator is to determine a relation of the input and output signals of a plant which is always fulfilled in the fault-free case and which may not be fulfilled in the case of a fault. Such a relation can be found by eliminating all unknowns from a contradictable constraint set which describes the nominal behavior of the plant. Since a subset of the constraints which describe the behavior of the plant in the fault-free case which is GMSO is typically contradictable under arbitrary excitation, GMSOs form the basis of the approach to the design of residual generators. This approach is:

1. Determine a constraint set $\mathcal{C}_{\mathrm{GMSO}} \subseteq \mathcal{C}$ in which the fault variable f_i appears.

2. Compute the unknown variables from the known variables under the assumption that the plant is fault-free. This generally requires $|\mathcal{C}_{\mathrm{GMSO}} \setminus c_k| = |\mathcal{C}_{\mathrm{GMSO}}| - 1$ constraints and a causal matching on the directed global structure graph of the GMSO.

3. Inject the resulting expressions for the unknown variables into the remaining constraint c_k. If the plant has I/O-pairs which match the nominal behavior - that is the assumption that was used to compute this expression - the resulting relation is always satisfied.

4. Bring all terms in this relation to one side. This results in a residual generator.

Similarly, it is possible to determine global residual generators if the remaining constraint is an inequality. If the unknown variables are computed from the known ones and all terms are brought to one side of the inequality, it becomes

$$0 > g(\boldsymbol{u}(t), \boldsymbol{y}(t)). \tag{4.87}$$

A residual is then

$$r(t) = \begin{cases} 0, & \text{if } g(\boldsymbol{u}(t), \boldsymbol{y}(t)) < 0, \\ g(\boldsymbol{u}(t), \boldsymbol{y}(t)), & \text{otherwise.} \end{cases} \tag{4.88}$$

If the elimination of the unknown variables from the GMSO is done according to the above procedure, the computation of a global residual generator requires the computation of the unknown variables in $\mathcal{X} \cap \mathrm{var}^G(\mathcal{C}_{\mathrm{GMSO}})$. Two properties guarantee that this is possible:

1. Any unknown variable which appears in $\mathcal{C}_{\mathrm{GMSO}}$ can be computed uniquely from a constraint $c \in \mathcal{C}_{\mathrm{GMSO}}$, known variables and previously computed unknown variables.

2. For computing an unknown variable, the knowledge of the value of this variable should not be necessary.

The consequences of both properties for the structural analysis are detailed in the following.

Existence of a causal matching. A main step in determining a global residual generator is to compute the unknown variables in a GMSO. For each unknown variable in this GMSO, one therefore needs to solve a constraint for this variable. This is not always uniquely possible. As a matter of fact, according to Definition 4.10, the information which constraints can be solved for which variables, is contained in the directed global structure graph. If a causal matching can be found on the unknown variables on the directed global structure graph of a GMSO, the value of each unknown variable appearing in the GMSO can be found. Of course, this requires the values of the other variables appearing in the GMSO to be known. If differential constraints are contained in the GMSO, the derivatives w.r.t. time may be necessary as well.

Because the acausal edges are marked with a "-1", it is possible to partition the part $M_\mathcal{X}$ of the incidence matrix of a directed global structure graph into a part M_c which contains only causal edges and a part $M_{\bar{c}}$ which contains both, causal and acausal edges. Such a partition is depicted in Fig. 4.15.

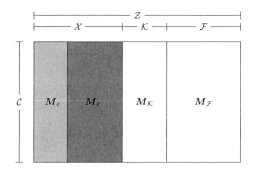

Figure 4.15: Partition of the incidence matrix into a strictly causal part M_c and a mixed causal and acausal part $M_{\bar{c}}$

Let there be a causal matching $\mathcal{M}_{\bar{c}}$ which is complete w.r.t. the unknown variables on the mixed part $M_{\bar{c}}$. Then the unknown variables and constraints which are not matched by this matching form a remaining part of the incidence matrix. This remaining incidence matrix can be found by removing the rows and columns which correspond to matched variables and constraints from the incidence matrix M. Let there further be a causal matching $\mathcal{M}_\mathrm{c}$ which is complete w.r.t. the unknown variables in the causal part M_c of the remaining part of the incidence matrix. Then the union

$$\mathcal{M} = \mathcal{M}_{\bar{c}} \cup \mathcal{M}_\mathrm{c} \tag{4.89}$$

of both matchings forms a causal matching $\mathcal{M}$ that is complete w.r.t. the unknown variables on the entire directed global structure graph.

The task to determine a causal matching is therefore split in two: determining a causal matching $\mathcal{M}_{\bar{c}}$ on the mixed part and determining a causal matching $\mathcal{M}_{c}$ on the causal part of the remaining incidence matrix.

For sound models there are usually only few acausal edges in a directed global structure graph and therefore the mixed part $\boldsymbol{M}_{\bar{c}}$ is small. For this reason, finding causal matchings on it can be done by checking all possible matchings which are complete w.r.t. the unknown variables for causality.

Because the causal part of the remaining incidence matrix only contains causal edges, an arbitrary matching found on it is causal. It is therefore sufficient to find any matching that is complete w.r.t. the unknown variables on the causal part $\boldsymbol{M}_{c}$ of the remaining incidence matrix in order to find a causal matching. This can be done with the help of the Dulmage-Mendelsohn-Decomposition.

The following algorithm sums up the approach to determine a causal matching on a directed global structure graph.

Algorithm 2 [DetermineCausalMatching]:

Input: The incidence matrix of a directed structure graph G.

Step 1: Permutate the incidence matrix $\boldsymbol{M}$ of G so that there is a strictly causal part $\boldsymbol{M}_{c}$ and a mixed causal part $\boldsymbol{M}_{\bar{c}}$.

Step 2: Determine causal matchings $\mathcal{M}_{\bar{c}}$ on the mixed causal part of the incidence matrix by trial and error. For each causal matching do:

Step 3: Use the Dulmage-Mendelsohn-Decomposition on the unmatched causal part of the remaining incidence matrix to check whether there is a causal matching and determine the causal matching $\mathcal{M}_{c}$. If there is a causal matching, there is a causal matching on the entire directed structure graph.

Step 4: Merge the matchings found in Step 2 and Step 3: $\mathcal{M} = \mathcal{M}_{c} \cup \mathcal{M}_{\bar{c}}$

Result: A causal matching $\mathcal{M}$, provided that Step 2 and Step 3 are successful.

Absence of cycles in the matching. The existence of a causal matching does guarantee that one can compute the value of an unknown variable from a constraint and the value of the other variables appearing in this constraint. However, it does not guarantee that one can compute the unknown variables in a constraint set from the known variables. This is because the order of computation defined by a matching may contain loops: Obviously, computing a variable z_1 from a variable z_2, which itself is computed from variable z_1 can be possible – but it is not guaranteed to be possible. Also, if differential constraints appear in a loop, the solution for the unknown variables is only unique if the initial conditions are known. However, in this work it is assumed that loops can be solved using symbolic manipulation and that the initial conditions are known.

In the following, the approach to determine global residual generators is carried out for Plant A.

Example 4.7 (Global residual generator for Plant A). *The algorithm* DETER-
MINECAUSALMATCHING *yields that there is a causal matching in the directed global
structure graphs of the four GMSOs obtained in Example 4.5. Solving the constraint sets*
$C_{\mathrm{GMSO},1}$, $C_{\mathrm{GMSO},2}$, $C_{\mathrm{GMSO},3}$ *and* $C_{\mathrm{GMSO},4}$ *under the assumption that the plant is fault-free,
that is* $f_i = 0$ *and injecting the result into the remaining constraint, one obtains the
following global residuals:*

$$r_1(t) = K_{\mathrm{u}}u(t) - \frac{1}{K_{\mathrm{y}1}}\left(\frac{y_1(t)}{T} - \frac{\mathrm{d}}{\mathrm{d}t}y_1(t)\right) \tag{4.90}$$

$$r_2(t) = K_{\mathrm{u}}u(t) - \frac{1}{T}\left(\frac{y_2(t)}{K_{\mathrm{y}2}K_{\mathrm{M}}K_{\mathrm{u}}u(t)}\right) - \frac{\mathrm{d}}{\mathrm{d}t}\left(\frac{y_2(t)}{K_{\mathrm{y}2}K_{\mathrm{M}}K_{\mathrm{u}}u(t)}\right) \tag{4.91}$$

$$r_3(t) = y_2(t) - \frac{K_{\mathrm{y}2}K_{\mathrm{M}}K_{\mathrm{u}}}{K_{\mathrm{y}1}}u(t)y_1(t) \tag{4.92}$$

$$r_4(t) = y_2(t) - K_{\mathrm{y}2}K_{\mathrm{M}}\frac{y_1(t)}{K_{\mathrm{y}1}}\left(\frac{y_1(t)}{TK_{\mathrm{y}1}} + \frac{\mathrm{d}}{\mathrm{d}t}\frac{y_1(t)}{K_{\mathrm{y}1}}\right) \tag{4.93}$$

The analytical proof that the signals $r_1(t)$, $r_2(t)$, $r_3(t)$ *and* $r_4(t)$ *are in fact residuals can
be done by injecting the input and output signals of the fault-free and the faulty plant into
the above expressions and showing that the properties of global residuals hold.*

In this section, an approach to determine global residual generators was presented. It
consists of determining GMSOs for which there is a causal matching which is complete
w.r.t. unknown variables on the directed global structure graph. Provided that cycles
in the matching can be solved using symbolic manipulation, it is possible to determine
global residual generators by eliminating the unknown variables from the GMSO.

4.8   Global Structural Detectability and Discriminability

In this section, the idea of global structural detectability and discriminability is presented.
The link is drawn between these concepts, the concepts of structural detectability and
isolability in the literature and the concept of I/O-detectability and I/O-discriminability
used in this thesis.

All conditions which are necessary in order to be able to compute a global residual
generator according to the approach presented in Section 4.7 can be verified by means
of the directed global structure graph. This motivates the idea to use these conditions
and the directed global structure graph to find out which faults may be detected or
discriminated by global residual generators obtained with the presented approach. This
leads to the properties in the following definition.

Definition 4.12 (Global structural diagnosabilty properties).

Global structural detectability. *A fault f_i is called globally structurally detectable, if there is a GMSO $\mathcal{C}_{\mathrm{GMSO}} \subseteq \mathcal{C}$ with a causal matching of the unknown variables $\mathcal{X} \cap \mathrm{var}^G(\mathcal{C}_{\mathrm{GMSO}})$ on the directed global structure graph G and the property*

$$f_i \in \mathrm{var}^G(\mathcal{C}_{\mathrm{GMSO}}) \tag{4.94}$$

holds.

Global structural discriminability. *Two faults f_i and f_j are called globally structurally discriminable, if there is a GMSO $\mathcal{C}_{\mathrm{GMSO}}$ with a causal matching of the unknown variables $\mathcal{X} \cap \mathrm{var}^G(\mathcal{C}_{\mathrm{GMSO}})$ on the directed global structure graph G and the property*

$$f_i \in \mathrm{var}^G(\mathcal{C}_{\mathrm{GMSO}}) \ \wedge \ f_j \notin \mathrm{var}^G(\mathcal{C}_{\mathrm{GMSO}}) \tag{4.95}$$

holds.

Both properties can be verified with the help of the global signature matrix $\boldsymbol{S}$. This matrix describes which faults may be the reason for a nonzero global residual. In the global signature matrix, fault variables correspond to columns. Residuals (or the GMSOs from which they are obtained) correspond to rows. The element $s_{k,i}$ of the matrix $\boldsymbol{S}$ is "1", if the fault variable f_i appears in the constraint set $\mathcal{C}_{\mathrm{GMSO},k}$ and 0 otherwise:

$$s_{k,i} = \begin{cases} 1, & \text{if } f_i \in \mathrm{var}^G(\mathcal{C}_{\mathrm{GMSO},k}) \\ 0 & \text{otherwise.} \end{cases} \tag{4.96}$$

The global signature matrix can be computed by applying the algorithm FINDMSO to the constraint set $\mathcal{C}$ in the analytical model using the directed global structure graph. Then, for the resulting GMSOs the existence of a causal matching is verified with the algorithm DETERMINECAUSALMATCHING. Faults which are not globally structurally detectable have an empty column in the global signature matrix $\boldsymbol{S}$. Two faults with the same entries in their columns are not globally structurally discriminable.

The difference between the global structural detectability and discriminability according to Definition 4.12 and the similar definition of structural detectability and isolability according to [40] is the following: Definition 4.12 requires a causal matching, whereas [40] doesn't use the concept of causality presented in this thesis. Therefore, structural detectability according to [40] may result in a fault being structurally detectable, although it is not possible to determine a global residual generator from the corresponding GMSOs. This limitation is solved by Definition 4.12. Similar reasoning holds for the structural discriminability / isolability.

Example 4.8 (Global signature matrix of Plant A). *With the GMSOs from Example 4.5 and the existence of a causal matching on these GMSOs known from Example 4.7, one obtains the global signature matrix in Tab. 4.4 for Plant A.*

Table 4.4: Global signature matrix of Plant A

S	f_1	f_2	f_3	f_4	f_5
$C_{\mathrm{GMSO},1}$	1	1	1	0	0
$C_{\mathrm{GMSO},2}$	1	1	0	1	1
$C_{\mathrm{GMSO},3}$	1	0	1	1	1
$C_{\mathrm{GMSO},4}$	0	1	1	1	1

The global signature matrix S reveals that all faults are globally structurally detectable, because there is at least one entry in each column. However, the faults f_4 and f_5 are not globally structurally discriminable since there is no C_{GMSO} which contains f_4, but not f_5 and vice versa.

Link between global structural diagnosability and I/O-diagnosability properties.
In the following, the relationship between global structural detectability and I/O-detectability is elaborated. The reasoning is based on the idea that GMSOs are typically contradictable under arbitrary excitation. The argumentation can be carried over to the relationship between global structural discriminability and I/O-discriminability.

Although global minimal structural overconstrainedness is neither necessary nor sufficient for the contradictability of a constraint set, the following train of thought holds for GMSOs: Typically, in a GMSO for which there is a causal matching that is complete w.r.t. the unknown variables on the directed global structure graph, the value of each unknown variable can be determined uniquely. For this purpose, the constraint to which an unknown variable is matched and the values of the other variables are used. This is possible, independently of the input signal $u(t)$. If all unknown variables are determined in this way and the result is injected into the remaining constraint, a consistency relation is obtained. Typically, for an arbitrary input signal $u(t)$, it is possible to choose a signal $y(t)$ such that if both are injected into the consistency relation, a contradiction results. Note that this is usually possible, independent of the nature of the input signal $u(t)$. One can therefore conclude that GMSOs are typically contradictable under arbitrary excitation.

According to Theorem 4.3, a fault f_i is I/O-detectable if there is an input signal $u(t)$ such that there is a constraint set $C_{\mathrm{U},f_i} \subseteq C_{f_i}$ which is contradictable and the output of the fault-free plant contradicts this constraint set.

Definition 4.12 states that if a fault f_i is globally structurally detectable, there is a GMSO in which the fault f_i appears. The above reasoning showed that GMSOs are typically contradictable under arbitrary excitation. They therefore fulfill the first condition for I/O-detectability. Usually, $y_0(t)$ contradicts the GMSOs of C_{f_i} in which f_i appears. This fulfills the second condition for I/O-detectability. For this reason, global structural detectability typically entails I/O-detectability.

A GMSO being contradictable means that one can find a signal $y(t)$ which leads to the contradiction of the GMSO. This is the general case. However, in some situations the consistency relation derived from the GMSO holds for arbitrary signals $y(t)$ – for instance, if constraints are inequalities which hold for arbitrary signals $\boldsymbol{y}(t)$. In that case, the GMSO is *not* contradictable.

The above reasoning relates global structural detectability and I/O-detectability: Global structural detectability usually results in I/O-detectability, but it is not guaranteed to do so. The same holds for global structural discriminability.

The definition of the I/O-detectability is satisfied if there is at least one input signal for which the output of the faulty plant is different from the output of the nominal plant, c.f. Theorem 4.1. However, this is usually the case for *most* of the input signals. Therefore, the output signals of the faulty plant typically contradict the GMSO and a residual generator obtained from the GMSO indicates the presence of a fault. For this reason, if there is a GMSO in which a fault variable appears, there is typically a residual generator which indicates a fault if it is present. It is this "typical" which often leads to the interpretation of a nonzero residual indicating a fault-free plant, which does in fact not match the definition of the residual.

This section discussed the concept of global structural detectability and discriminability and their relation to the structural detectability and isolability in the literature. The link between the structural diagnosability properties, based on GMSOs and the I/O-detectability and I/O-discriminability, lies in the contradictability of GMSOs which is typically given under an arbitrary excitation.

4.9   Limits of Global Structural Analysis

This section details the major limit of global structural analysis: global structural detectability and global structural discriminability are neither necessary nor sufficient for I/O-detectability and I/O-discriminability. The reasons for this motivate the approach to a more precise analysis in Chapter 5.

Global structural analysis is a powerful tool to roughly determine diagnosability properties of dynamical systems. It also provides a basis for the computation of global residual generators which realize consistency tests. As pointed out in the previous section, the core of the relationship between the global structural diagnosability properties and the I/O-diagnosability properties is that GMSOs are typically contradictable. One would have wanted the structural analysis to give necessary or sufficient conditions for the I/O-diagnosability, but this is not the case:

On one hand, although global structural detectability of a fault allows to determine global residual generators which are typically sensitive to a fault, it does not guarantee that the residual provided by such a residual generator becomes nonzero if the fault is present. The same observation holds for global structural discriminability: Typically, structural discriminability of two faults allows to determine a residual generator which is sensitive to one fault, but not to the other. However, if a global residual generator was found for which global structural analysis allowed to conclude that it is not sensitive to one fault, it is not guaranteed that this residual generator is sensitive to the other fault.

On the other hand, it is possible that a fault is not globally structurally detectable although a specific input signal allows to determine whether the fault is present and, therefore, the fault is I/O-detectable. Similarly, faults which are not globally structurally discriminable may be distinguished with a specific excitation and analysis of the plant's output signals. In order to illustrate these effects, a number of examples are given here:

Counterexample 1. Let the constraint set $\mathcal{C} = \{c_1, c_2\}$ describe a static SISO-system with the unknown but deterministic variable x_2:

$$c_1 : \quad 0 = y(t) - x_1(t)x_2(t) + f_1 \tag{4.97}$$

$$c_2 : \quad 0 = u(t) - x_1(t) \tag{4.98}$$

The analysis of the global structure graph of $\mathcal{C}$ yields that there is no subset of $\mathcal{C}$ that is GMSO. However, the constraint set $\mathcal{C}_{f_1}$ is contradictable because if $u(t) = 0$, any output signal $y(t) \neq -f_1$ leads to a contradiction with the constraint set $\mathcal{C}_{f_1}$. Obviously, the existence of a subset which is GMSO is not necessary for the contradictability of a constraint set. Also, for the same input signal $u(t) = 0$, the plant answers with output signals which are different in the fault-free case ($y_0(t) = 0$) and in the faulty case ($y_{f_1}(t) = -f_1$). This allows with Theorem 4.1 to conclude that the fault f_1 is I/O-detectable. Hence, the existence of a GMSO in which the fault variable f_i appears is not necessary for the I/O-detectability of the fault f_i.

Counterexample 2. Let the constraint set $\mathcal{C} = \{c_1, c_2, c_3\}$ with

$$c_1 : \quad 0 = u(t) - x_1(t) + f_1 \tag{4.99}$$

$$c_2 : \quad 0 = x_2(t) - x_1(t) \tag{4.100}$$

$$c_3 : \quad 0 = y(t) - x_2(t) + f_1 \tag{4.101}$$

describe a static SISO-system which may be subject to the fault f_1. The analysis of the system's global structure graph G yields that the constraint set $\mathcal{C}$ is GMSO and that the causal-matching-condition is satisfied. Hence, the fault f_1 is globally structurally detectable. The constraint set $\mathcal{C}$ is also contradictable under an arbitrary excitation $u(t)$: Solving the constraint set $\mathcal{C}_{f_i}$ for the unknown variables in it yields $x_1(t) = u(t) + f_1$ and $x_2(t) = u(t) + f_1$. Since $x_2(t) = y(t) + f_1$ also holds, it is possible to choose $y(t) \neq u(t)$ which leads to contradiction with the constraint set $\mathcal{C}$. However, although the fault variable f_1 appears in the GMSO, this fault does not lead to such an output signal. This is because solving the constraint set $\mathcal{C}_{f_1}$, one obtains $y(t) = u(t)$, independently of the value of f_1. Hence, a fault may not be I/O-detectable although it is globally structurally detectable. Obviously, in this rather simple example, one can directly perceive that a change in the fault variable f_1 does, in fact, not change the behavior of the plant. In larger and more complex constraint sets, a similar situation, in which a fault actually has no impact, cannot be guaranteed to be noticed prior to the structural analysis.

Counterexample 3. Let the constraint set $\mathcal{C} = \{c_1, c_2, c_3\}$ with

$$c_1 : \quad 0 = x_1(t) + x_2(t) - y(t) \tag{4.102}$$

$$c_2 : \quad 0 = ax_1(t) + bx_2(t) - y(t) \tag{4.103}$$

$$c_3 : \quad 0 = x_1(t) - u(t) + f_1 \tag{4.104}$$

describe a static SISO-system. The result of the analysis of its directed global structure graph G is that C is GMSO and that the causal-matching-condition is satisfied. The fault f_1 is therefore globally structurally detectable. However, the constraint set is not contradictable[3] for $a = b = 1$. Hence, global structural overconstrainedness is not sufficient for contradictability and therefore not sufficient for I/O-detectability. Although this example violates the assumption that the models are elimination-minimal in the sense of Section 3.3.4 and the condition that constraints are not allowed to be linear combinations[4] given in [22], it illustrates one main property of the global structural analysis: The result of the global structural analysis is the same, regardless of the parameters in the constraints. In contrast to this, the contradictability of a constraint set and the I/O-diagnosability properties are not independent of the parameters. The assumption that the models are elimination-minimal in Section 3.3.4 and the condition on linear combinations of the constraints try to solve these issues. However, these conditions do not overcome the problem of the dependency on parameters completely.

Similar counterexamples can be found for the relationship between structural discriminability and I/O-discriminability. In order to illustrate the effect described in the counterexamples in the I/O-plane, in the following a plant is considered which may be subject to five different faults. The behaviors of the plant in the fault-free and in the faulty cases are depicted in Fig. 4.16. The I/O-pairs which are consistent with the nominal behavior $\mathcal{B}_0$ are in dark gray and the behaviors $\mathcal{B}_{f_1} - \mathcal{B}_{f_5}$ in the case of the faults $f_1 - f_5$ in middle gray. Assume that the global structural analysis resulted in two GMSOs $C_{\text{GMSO},1}$ and $C_{\text{GMSO},2}$ which allow to determine two global residual generators. The I/O-pairs which are consistent with the GMSOs are depicted in light gray. The signature matrix which corresponds to the GMSOs is given in Tab. 4.5.

Table 4.5: Signature matrix corresponding to the behaviors in Fig. 4.16

S	f_1	f_2	f_3	f_4	f_5
$C_{\text{GMSO},1}$		1	1	1	
$C_{\text{GMSO},2}$	1		1	1	

Since the faulty behaviors and the nominal behavior do not intersect pairwisely, all faults are I/O-detectable and I/O-discriminable. However, fault f_5 is not structurally detectable because it has an empty column in the signature matrix. The behavior $\mathcal{B}_{f_5}$ has only elements which are contained in the behaviors associated with the two residual generators found with the global structural analysis. Therefore, no I/O-pair which results in the case of fault f_5 is inconsistent with the behaviors $\mathcal{B}(C_{\text{GMSO},1})$ and $\mathcal{B}(C_{\text{GMSO},2})$. The global residual generators can therefore never detect the fault f_5. This is the case, although fault f_5 is obviously I/O-detectable: The behavior in the case of fault f_5 is not the same as the nominal behavior.

---

[3]As a matter of fact, the fault f_1 is not I/O-detectable for $a = b = 1$.

[4]Although usually sound models of physical systems do not contain constraints that are linear combinations of other constraints, there is no guarantee for that.

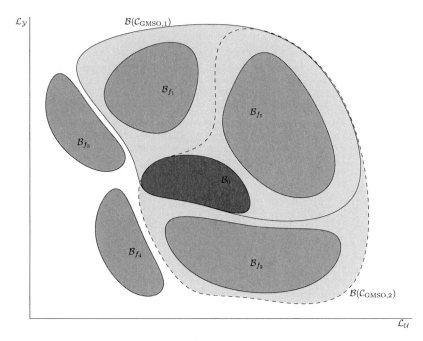

Figure 4.16: Global structural vs. I/O-diagnosability properties in the I/O-plane

Also, fault f_3 and fault f_4 form a group of faults which are globally structurally not discriminable since the columns which correspond to the faults in the signature matrix have identical entries. This is the case, although they are I/O-discriminable because the corresponding behaviors $\mathcal{B}_{f_3}$ and $\mathcal{B}_{f_4}$ are not the same.

The above examples show that

a)   global structural overconstraindness does not guarantee contradictability and

b)   the contradiction of a contradictable constraint subset of the faulty model by the output signals of the fault-free plant is not guaranteed by the subset's contradictability.

For these reasons, global structural analysis fails at providing necessary and sufficient conditions in order to exactly infer on the I/O-detectability and I/O-discriminability.

However, global structural diagnosability properties are practical properties because they result in "typically"-statements for the I/O-detectability and I/O-discriminability. They therefore allow to roughly infer on the I/O-diagnosability properties. Also, they form the basis for an approach to the design of global residual generators. Particularly in a passive diagnosis setting, in which the diagnosis has to be performed without influencing the input signals, the global structural analysis is a powerful tool to treat large-scale nonlinear systems.

One reason for the gap between the results of the global structural analysis and the actual I/O-diagnosability properties is the effect of specific input signals which is not captured by the global structure graph: Counterexample 1 showed that a constraint set may be contradictable under a specific excitation although it is not GMSO.

In contrast to the passive diagnosis setting, active diagnosis aims at applying specific input signals to the plant for diagnostic purposes. Since these input signals may be of arbitrary form in automatic tests, they can exploit a part of the gap between the results obtained with global structural analysis and I/O-diagnosabilty properties.

Analyzing and exploiting the cause for this gap may help determining input signals and means to infer on the fault-state of the plant under the specific excitation. For this reason, the tools from global structural analysis are applied in Chapter 5 to models of plants which are assumed to have been brought to specific operating regions by appropriate input signals. In Chapter 6, a graph-based approach to determine signals which allow to steer a plant into such operating regions is presented.

In this section, counterexamples were used to present the limits of global structural analysis. The reasons for these limits were analyzed.

4.10 Summary

The chapter started with the description of consistency tests for diagnosis. These tests determine whether an I/O-pair observed at the plant is contained in its nominal behavior. Specific signals which are called global residuals are used for this purpose. Two properties of dynamical systems which may be subject to faults were introduced:

- I/O-detectability which answers the question whether a specific fault changes the behavior of the system and may therefore be detected on the basis of the system-behavior, and

- I/O-discriminability which is the property that two faults change the behavior of the plant in a specific way so that they can be distinguished on the basis of the behavior regardless of the fault-magnitudes.

It was shown that both properties are related to the contradictability of the constraint set describing the plant and subsets thereof. A structural representation of constraint sets which allows to determine constraint subsets that are typically contradictable under arbitrary excitation was introduced. This representation is a bipartite graph. It contains the information which variables interact with each other via which constraints in an arbitrary operating region. The graph was therefore called global structure graph. The global structure graph was extended by the information whether a specific variable can be computed from a constraint. This resulted in the directed global structure graph. Along with two algorithms, the directed global structure graph allowed to determine constraint sets from which global residual generators can be obtained. The directed global structure graph was also used for the evaluation of structural diagnosability properties of the faults. Although these are useful properties for passive diagnosis, it was shown that these properties are neither necessary nor sufficient conditions for the corresponding I/O-diagnosability properties.

Since the global residual generators exhibit the specific property to be zero in the fault-free case regardless of the system excitation, they are very well-suited for the supervision and onboard-diagnosis of technical systems. However, the global residuals have some disadvantages with respect to fault discrimination. This is mainly due to the fact that the number of global residual generators found with global structural analysis is limited.

In a setting where the system does not need to fulfill its original functional goal (completely) it can be excited with specific input signals. These input signals can steer the plant into specific operating regions which may allow to find more specific consistency checks, thus improving fault discrimination.

The next chapter is concerned with the development of this idea.

Chapter 5

Analysis of Local Structure Graphs

5.1 Locality of Couplings and its Consequence for Diagnosis

The previous chapter closed with the observation that a fault may be I/O-detectable, although it is not globally structurally detectable. The same holds for the I/O-discriminability of two faults which can be given, although the faults are not globally structurally discriminable. One reason for this is that the global structure graph exaggerates the couplings in the plant: If a coupling between a fault and a GMSO is weak or not present in a specific operating region, the global structure graph still contains an edge between the constraints in the GMSO and the fault variable. However, if the system is known to be in that operating region, the coupling represented by the edge vanishes. Then the edge can be dropped from the global structure graph, thus forming a new graph. The results obtained from the structural analysis conducted on this new graph will result in the GMSO not being sensitive to the fault. The results of the analysis of the new graph is therefore closer to the I/O-diagnosability properties than the result of the analysis of the global structure graph. This observation motivates the approach taken in this chapter:

1. Search for operating points in which couplings vanish.

2. Use a new graph which takes into account the vanished couplings for the analysis of diagnosability properties and the construction of consistency tests.

In Fig. 5.1, the relation between three variables defined by a constraint c is depicted. For the specific values $\bar{z}_1$ and $\bar{z}_2$ of the variables z_1 and z_2, the influence of the third variable z_3 vanishes. If the variables z_1 and z_2 are ensured to take on these values, the coupling between z_3 and the variables z_1 and z_2 is not present. One can then drop the edge between the variable-vertex z_3 and the constraint-vertex c from the global structure graph.

Such a structural breakdown in specific operating regions motivates the definition of the first contribution of this chapter, the directed local structure graph. This graph describes the couplings among the variables in a constraint set, if some variables are restricted to specific operating regions. This is done by dropping edges from the global structure graph and adding constraints which describe the operating region.

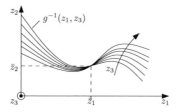

Figure 5.1: Values consistent with a constraint, projected into the $z_1 z_2$-plane

The second contribution is the local residual. This is a signal which has the same properties as a global residual under the condition that the plant is in a specific operating region. Whereas the global residual generator allows the fault detection for arbitrary input signals, the local residual has this property only for specific input signals: namely the signals which steer the plant into the specific operating regions.

The third contribution is an approach to determine local residual generators. It is based on the directed local structure graph which holds in a specific operating regions and makes use of the algorithms in [76]. Provided that steering the system into the specific operating region, in which the local structure graph holds, is successful, the local residual generators have a better ability to detect and to discriminate faults than the global residual generators.

In Section 5.2, the notion of an operating region which can be used in the modeling framework known from Chapter 3 is described. A method to determine operating regions in which the coupling between a variable and a constraint vanishes is introduced in Section 5.3. These operating regions are used to define the directed local structure graph in Section 5.4. Consequences of the analysis of the local structure graph for the design of diagnostic tests, including the introduction of the local residual, are given in Section 5.5.

5.2   Operating Region

Starting from the definition of an operating point and an operating region the notation used for their representation in this thesis is given.

The basic principle of the approach to the solution of the diagnostic problem described in Section 2.7 is steering the plant in a specific manner. This demands a formal description of the goal of such a kind of steering. The modeling approach of analytical models used in this thesis (c.f. Chapter 3) motivates the idea to describe the goal of the steering in a similar way. The variables are the part of an analytical model that represent the current state[1] of the plant. A tuple of values for the variables which is consistent with the constraints in the analytical model of a static plant represents a physically reasonable state of the plant.

---

[1]Here, the expression 'state' is not meant in the sense of a state-vector of the state-space representation of a dynamical system.

Since the plant may be a dynamical system, its analytical model may contain differential equations or differentiation-operations as constraints. In that case, considering only the values of the variables at a specific instant of time is not sufficient for the definition of an operating point. This is because at a specific instant of time, the value of a variable and the value of its derivative w.r.t. time cannot be proven to be inconsistent with the differential relation linking these two variables. Defining an operating point solely on the basis of values of the variables can therefore lead to tuples of values which do not describe a reasonable physical state of a plant if its analytical model contains differential constraints.

One approach to overcome this problem is to define an operating point as a tuple of values for the variables and their derivatives w.r.t. time at a specific instant of time. This leads to the following definition:

Definition 5.1 (Operating point and region). *An* operating point *is a tuple of values for the variables in an analytical model and their derivatives w.r.t. time that does not contradict the analytical model. An* operating region *is a set of operating points.*

According to the above definition, any element of an operating region is a tuple of values for the variables and their derivatives w.r.t. time which satisfies the constraint set $\mathcal{C}$ in an analytical model. The set of all possible operating points – which is the same as the largest possible operating region – is therefore defined by the constraint set $\mathcal{C}$. A subset of this largest operating region can be described as a further restriction of the variables and their derivatives w.r.t. time. One way to express such an additional restriction is the introduction of a set $\mathcal{C}_w$ of additional constraints which the variables and their derivatives need to satisfy. In that way, an operating region is defined by the set of tuples which do not contradict the constraint set

$$\mathcal{C} \cup \mathcal{C}_w. \tag{5.1}$$

An operating region contains at least one operating point. Hence, a constraint set $\mathcal{C}_w$ is not an operating region if there is no tuple of values and derivatives w.r.t. time which satisfies the constraint set $\mathcal{C} \cup \mathcal{C}_w$. For ease of readability, in the following only the additional constraints $\mathcal{C}_w$ are used when an operating region is described. An example of a constraint set defining an operating region is given in Example 5.1:

Example 5.1 (Operating region for Plant A). *An example of an operating region of Plant A which restricts the variable x_3 is*

$$\mathcal{C}_w = \{c: \quad 0 = x_3\}. \tag{5.2}$$

In the simplest case, only known variables appear in the constraints in the set $\mathcal{C}_w$. In that case, the operating region is a simple restriction of the I/O-pairs in the I/O-plane. This type of restrictions of the I/O-pairs does not depend upon couplings in the plant.

The representation of such an operating region in the I/O-plane is depicted in Fig. 5.2. Note also that in this case, the I/O-pairs allowed by the constraint set $\mathcal{C}_w$ are independent of the fault-state of this plant.

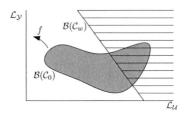

Figure 5.2: Operating region: restriction in the I/O-plane independent of the couplings in the plant and its fault-state

One type of such an operating region is a control law $u(t) = f(y(t))$ since it directly links the inputs to the outputs of the plant.

However, in the general case unknown variables appear in the description $\mathcal{C}_w$ of the operating region. If one still wants to describe the operating region as a restriction in the I/O-plane, a restriction of unknown variables needs to be mapped to a restriction of known variables. This is possible by using a part of the constraints in the analytical model of the plant because they provide the relation between known and unknown variables.

For the plants considered in this thesis, there are two properties which lead to the fact that the restriction in the I/O-plane which results from the description of an operating region $\mathcal{C}_w$ and model information $\mathcal{C}$ is not unique:

First, to obtain a restriction in the I/O-plane, it is possible to use different parts of the plant model to map constraints on unknown variables in $\mathcal{C}_w$ to constraints on known variables only. These different parts of the model correspond to different subsets of the constraint set $\mathcal{C}$ in the analytical model. For this reason, an operating region $\mathcal{C}_w$ can be described by different restrictions in the I/O-plane if the constraint set $\mathcal{C}_w$ contains unknown variables.

Second, the fault-state of a plant typically influences the relation between known variables and unknown variables. Therefore, even if the same model information, that is, the same subset of $\mathcal{C}$, is used to map the constraint set $\mathcal{C}_w$ to a restriction of the known variables, the resulting restriction in the I/O-plane is not necessarily fixed. This is due to the potentially faulty nature of the plant. If a part of the plant model that is used to map the constraints on unknown variables to a constraint on known variables may be subject to a fault, the presence of this fault may alter the resulting restriction in the I/O-plane.

For these reasons, if the constraint set $\mathcal{C}_w$ is mapped to a restriction in the I/O-plane, it is necessary to state which model information from $\mathcal{C}$ and which fault-state was used. In Fig. 5.3 the effect of using the model information $\mathcal{C}_{U,f_i} \subset \mathcal{C}$ to map an operating point to a restriction in the I/O-plane is depicted. In contrast to the situation illustrated in Fig. 5.2, not only the plant behavior, but also the restriction on the I/O-pairs depends on the magnitude of the fault.

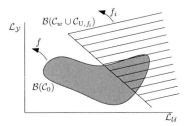

Figure 5.3: Operating region: restriction in the I/O-plane dependent on the couplings in the plant and its fault-state

The above reasoning implies that operating regions described by constraint sets which contain unknown variables are difficult to use in an interpretation on the basis of I/O-pairs. It is therefore difficult to infer on the presence of such an operating region at a potentially faulty plant. However, the ideas in this reasoning are used in Section 6.4 for the definition of a signal that allows to determine the presence of a specific operating region. This signal is used in tests that depend upon the operating region of the plant.

In this section, the definition of operating points and operating regions was given. A notation for operating regions was introduced and illustrated in the I/O-plane. Finding operating regions $\mathcal{C}_w$ which are advantageous for inferring on the I/O-diagnosability properties and the fault-state of a plant is the subject of the next sections.

5.3 Additional Structural Properties Valid in Specific Operating Regions

This section introduces the idea of structural changes due to the restriction of the plant to specific operating regions.

Typically, physical systems have the property that in some operating regions the couplings among two variables are strong, e.g. a change in the value of one variable influences another variable significantly. In some other operating regions, the coupling among the same variables is weak, e.g. a change in the value of one variable barely affects the value of the other variable. In some operating regions, the influence of a variable z on the manifold defined by the constraint c may even vanish completely although the variable z appears in the constraint c. This situation is depicted in Fig. 5.4 and Fig. 5.5: The graph in Fig. 5.4 represents the relation between three variables, z_1, z_2 and z_3 formed by the constraint c

$$c: \quad 0 = g(z_1, z_2, z_3) \tag{5.3}$$

as a graph $z_2 = g^{-1}(z_1, z_3)$ for different values of z_3.

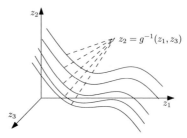

Figure 5.4: Values consistent with the constraint c, 3D-plot

In Fig. 5.5, the same graph is projected into the $z_1 z_2$-plane. This projection reveals that for specific values $\bar{z}_1$ and $\bar{z}_2$ for the two variables z_1 and z_2, the relation between z_1 and z_2 is independent of the third variable z_3.

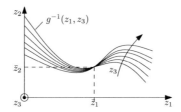

Figure 5.5: Values consistent with the constraint c, projection into the $z_1 z_2$-plane

The variables z_1 and z_2 having specific values can be interpreted as an operating region $\mathcal{C}_w = \{c_1, c_2\}$ with the constraints

$$c_1 : 0 = z_1 - \bar{z}_1 \tag{5.4}$$
$$c_2 : 0 = z_2 - \bar{z}_2. \tag{5.5}$$

Therefore, the variable z_3 does not have any impact on the constraint c, provided that the operating region $\mathcal{C}_w$ is present. For illustrative reasons, this example has been chosen such that the resulting operating region is rather simple. Obviously, for other constraints, the operating region derived from the projection may be more complex. More generally, in particular operating regions, the influence of a variable z_j on the manifold defined by the constraint c_i may vanish. This can be the case, although the variable z_j appears in the constraint c_i and, therefore, there is an edge between the variable-vertex z_j and the constraint-vertex c_i in the global structure graph. The fact that edges in the global structure graph represent couplings among the variables, motivates the idea to drop an edge from the global structure graph if the corresponding coupling vanishes. This requires to describe, under which conditions a change in the value of one variable does not influence the solution of a constraint for the other variables. This is done in the following.

Let c_i be an algebraic constraint in which the variables $\mathcal{Z}_i = \{z_1, z_2, ..., z_r, ..., z_j, ..., z_k\}$ appear

$$c_i : \quad 0 = g_i(z_1, z_2, ..., z_r, ..., z_j, ..., z_k) \tag{5.6}$$

and let

$$z_r = g_i^{-1}(z_1, z_2, ..., z_{r-1}, z_{r+1}, ..., z_j, ..., z_k) \tag{5.7}$$

be the solution of the implicit function g_i for an arbitrary variable $z_r \in \mathcal{Z}_i$. A mathematical expression for the variable z_j not having any influence on the solution for z_r is

$$\frac{\mathrm{d}}{\mathrm{d}z_j} z_r = \frac{\mathrm{d}}{\mathrm{d}z_j} g_i^{-1}(z_1, z_2, ..., z_{r-1}, z_{r+1}, ..., z_j, ..., z_k) = 0 \ \forall \ z_j. \tag{5.8}$$

Instead of evaluating the above condition for all variables appearing in the constraint set $\mathcal{Z}_i \setminus z_j$ in order to obtain a condition on the other variables, the implicit function g_i can be investigated directly. The reason for this is that if the function g_i itself is invariant w.r.t. the variable z_j under certain conditions on the other variables, any rearrangement of g_i is invariant w.r.t. the variable z_j as well. This holds in particular for rearrangements that represent the solution of g_i for one of the variables in it. Therefore,

$$\frac{\mathrm{d}}{\mathrm{d}z_j} g_i(z_1, z_2, ...z_k) = 0 \ \forall \ z_j \tag{5.9}$$

can be evaluated instead of eqn. (5.8) in order to obtain a condition under which the variable z_j does not have an impact on the constraint c_i. If c_i is a differential equation, it is not sufficient to ensure that eqn. (5.9) holds, since derivatives w.r.t. time of the variables in $\mathcal{Z}_i$ appear in the function g_i as well. It is therefore necessary to extend eqn. (5.9) to

$$\frac{\mathrm{d}}{\mathrm{d}\left(\frac{\mathrm{d}^n}{dt^n} z_j\right)} g_i(z_1, z_2, ...z_k) = 0 \ \forall \ z_j, \ \forall \ n \in \mathbb{N}_0. \tag{5.10}$$

The above equation yields constraints under which neither the variable z_j nor any of its derivatives w.r.t. time influences the solution g_i^{-1} of the constraint c_i for another variable. As long as these conditions are met, a change of the value of the variable z_j does not have any effect on the constraint. This reasoning leads to the definition:

Definition 5.2 (Local ineffectiveness). *An edge e_{ij} between the vertices for the variable $z_j \in \mathcal{Z}_i$ and the constraint $c_i(\mathcal{Z}_i)$ in the global structure graph G is called locally ineffective under the condition $c_{\mathrm{Elim},ij}(\mathcal{Z}_p)$, if the condition*

$$c_{\mathrm{Elim},ij}(\mathcal{Z}_p) : \ 0 = c_{\mathrm{Elim},ij}(z_1^p, z_2^p, ...) = \frac{\mathrm{d}}{\mathrm{d}\left(\frac{\mathrm{d}^n}{dt^n} z_j\right)} g_i(z_1^p, z_2^p, ..., z_1^i, z_2^i, ...) \ \forall \ z_j \in \mathbb{R} \ \forall \ n \in \mathbb{N}_0$$

$$\text{with } \quad z_k^p \in \mathcal{Z}_p \quad \text{and} \quad z_k^p, z_k^i \in \mathcal{Z}_i \tag{5.11}$$

on the variables in the set $\mathcal{Z}_p \subseteq \mathcal{Z}_i$ holds. For the other values of the variables $\mathcal{Z}_p$, the edge is called locally effective.

Because of the condition $\mathcal{Z}_p \subseteq \mathcal{Z}_i$, no additional variables appear in $c_{\mathrm{Elim},ij}$. At most the variables which appear in the constraint c_i, that is the set $\mathcal{Z}_i$, appear in the constraint under which the edge e_{ij} is locally ineffective.

If there are only algebraic constraints, Definition 5.2 can be simplified to $n = 0$. If all dynamics in a constraint set are expressed by differential constraints of the form

$$d: \ 0 = \dot{z}_i(t) - \frac{\mathrm{d}}{\mathrm{d}t} z_i(t), \tag{5.12}$$

Definition 5.2 can be simplified to $n \in \{0, 1\}$. In the following example, three edges are investigated for local ineffectiveness.

Example 5.2 (Locally ineffective edge in Plant A). *The edges which are investigated for local ineffectiveness are e_1, e_2 and e_3 in the directed global structure graph of Plant A which is depicted in Fig. 5.6.*

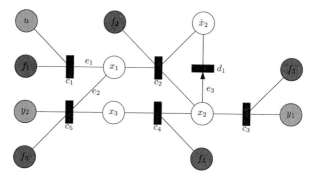

Figure 5.6: Edges in the directed global structure graph of Plant A which are investigated for local ineffectiveness

Investigation of edge e_1:
The application of the condition in Definition 5.2 to the edge e_1 yields:

$$c_{\mathrm{Elim},1} : 0 = \frac{\mathrm{d}}{\mathrm{d}\left(\frac{\mathrm{d}^n}{\mathrm{d}t^n} x_1(t)\right)} g_1\big(u(t), x_1(t), f_1\big)$$

$$c_{\mathrm{Elim},1} : 0 = \frac{\mathrm{d}}{\mathrm{d}\left(\frac{\mathrm{d}^n}{\mathrm{d}t^n} x_1(t)\right)} \big(x_1(t) - (1 + f_1) K_u u(t)\big)$$

$$c_{\mathrm{Elim},1} : 0 = 1 \ \ for \ \ n = 0$$

Since for arbitrary values of the other variables, the constraint $c_{\mathrm{Elim},1}$ may never be satisfied, the edge e_1 is locally effective in arbitrary operating regions. It is not necessary to investigate derivatives w.r.t. time of a higher order to obtain this result.

Investigation of edge e_2:
The application of the condition in Definition 5.2 to the edge e_2 yields:

$$c_{\text{Elim},2} : 0 = \frac{\mathrm{d}}{\mathrm{d}\left(\frac{\mathrm{d}^n}{\mathrm{d}t^n}x_1(t)\right)} g_5(y_2(t), x_1(t), x_3(t), f_5)$$

$$c_{\text{Elim},2} : 0 = \frac{\mathrm{d}}{\mathrm{d}\left(\frac{\mathrm{d}^n}{\mathrm{d}t^n}x_1(t)\right)} (y_2(t) - K_{y2}x_1(t)x_3(t) - f_5)$$

$$c_{\text{Elim},2} : 0 = K_{y2}x_3(t) \quad for \quad n = 0;$$

$$c_{\text{Elim},2} : 0 = 0 \quad for \quad n = 1;$$

Since all differential equations are expressed as differential constraints in this example, investigating $n = \{0,1\}$ is sufficient. Edge e_2 is locally ineffective under the condition $c_{\text{Elim},2} : 0 = x_3(t)$. Theoretically, since the constraint c_5 is an algebraic constraint, it would not have been necessary to investigate the condition for local ineffectiveness for $n > 0$. The result can be interpreted as follows: If $x_3(t) = 0$ holds, the relation between the variables x_3, y_2 and f_5 does not depend upon x_1.

Investigation of edge e_3:
The application of the condition in Definition 5.2 to the edge e_3 yields:

$$c_{\text{Elim},3} : 0 = \frac{\mathrm{d}}{\mathrm{d}\left(\frac{\mathrm{d}^n}{\mathrm{d}t^n}x_2(t)\right)} g_{d1}(\dot{x}_2(t), x_2(t))$$

$$c_{\text{Elim},3} : 0 = \frac{\mathrm{d}}{\mathrm{d}\left(\frac{\mathrm{d}^n}{\mathrm{d}t^n}x_2(t)\right)} \left(\dot{x}_2(t) - \frac{\mathrm{d}}{\mathrm{d}t}x_2(t)\right)$$

For $n = 0$, applying the theorem of Schwarz, c.f. [25], the above equation yields

$$c_{\text{Elim},3} : 0 = \frac{\mathrm{d}}{\mathrm{d}x_2(t)} \left(\dot{x}_2(t) - \frac{\mathrm{d}}{\mathrm{d}t}x_2(t)\right) = \left(\frac{\mathrm{d}}{\mathrm{d}x_2(t)}\dot{x}_2(t)\right) - \left(\frac{\mathrm{d}}{\mathrm{d}x_2(t)}\frac{\mathrm{d}}{\mathrm{d}t}x_2(t)\right)$$

$$= 0 - \frac{\mathrm{d}}{\mathrm{d}t}\frac{\mathrm{d}}{\mathrm{d}x_2(t)}x_2(t) = -\frac{\mathrm{d}}{\mathrm{d}t}1 = 0$$

and for $n = 1$

$$c_{\text{Elim},3} : 0 = \frac{\mathrm{d}}{\mathrm{d}\left(\frac{\mathrm{d}}{\mathrm{d}t}x_2(t)\right)} \left(\dot{x}_2(t) - \frac{\mathrm{d}}{\mathrm{d}t}x_2(t)\right) = \left(\frac{\mathrm{d}}{\mathrm{d}\left(\frac{\mathrm{d}}{\mathrm{d}t}x_2(t)\right)}\dot{x}_2(t)\right) - \left(\frac{\mathrm{d}}{\mathrm{d}\left(\frac{\mathrm{d}}{\mathrm{d}t}x_2(t)\right)}\frac{\mathrm{d}}{\mathrm{d}t}x_2(t)\right)$$

$$= 0 - 1 = -1$$

results. Since for arbitrary values of the variable $\dot{x}_2(t)$ the result for $n = 1$ is a contradiction, the edge e_3 is locally effective in arbitrary operating regions.

The result for the edge e_3 reveals the importance of considering the derivatives w.r.t. time when investigating the local ineffectiveness of edges in the global structure graph: If only the value and not the derivative w.r.t. time had been considered, the result of the analysis would have been that the edge e_3 is locally ineffective – although obviously, a change in the value of x_2 does affect $\dot{x}_2(t)$.

Using the concept of local ineffectiveness, it is also possible to determine the edges of the global structure graph in another way than determining which variables appear in which constraint: It is assumed that there is an edge between all variables and all constraints. In that way the bipartite graph has the maximum number of edges – the graph is complete under the bipartite condition. Then, each edge is investigated for local ineffectiveness. If an edge is found to be locally ineffective, regardless of the operating region – that is if $c_{\text{Elim},ij} : 0 = 0$ holds – the edge is eliminated from the structure graph. In that way, the global structure graph is obtained.

In this section, the concept of local ineffectiveness of edges in the global structure graph was introduced. This property describes the effect that some variables do not have any impact on a constraint if the other variables appearing in this constraint take on specific values – despite the fact that there is an edge between this variable and the constraint in the global structure graph. The local ineffectiveness of edges plays an important role in the definition of the local structure graph in the next section.

5.4 Directed Local Structure Graph

This section introduces the directed local structure graph. It is based on the observation of the structural breakdown in specific operating regions and the definition of local ineffectiveness. The directed local structure graph describes the couplings among the variables for a plant which is restricted to a specific operating region.

Let $c_{\text{Elim},ij}$ be a constraint that describes an operating region, in which the edge e_{ij} in the global structure graph G between the constraint-vertex c_i and the variable-vertex z_j is locally ineffective. If the plant is restricted to the operating region $\{c_{\text{Elim},ij}\}$, two consequences result for the analytical model of the plant:

First, the information on the operating region to which the plant is restricted needs to be incorporated into the analytical model. Since the operating region is given in the form of an additional constraint on the variables, this constraint can be added straightforwardly to the constraint set in the analytical model:

$$\mathcal{C} \cup \{c_{\text{Elim},ij}\} \tag{5.13}$$

Second, since the variable z_j does not influence the constraint c_i in the operating region defined by $c_{\text{Elim},ij}$, the constraint c_i can be expressed – possibly by injecting a rearrangement of $c_{\text{Elim},ij}$ into c_i – without the variable z_j. In this way, a new constraint $\tilde{c}_i$ with the property

$$z_j \notin \text{var}^{G(\{\tilde{c}_i\})}(\tilde{c}_i). \tag{5.14}$$

is obtained. This formula represents the structural breakdown when $G(\{\tilde{c}_i\})$ signifies the global structure graph of $\tilde{c}_i$.

The above steps result in the constraints of the analytical model in the operating region described by $c_{\mathrm{Elim},ij}$:

$$(\mathcal{C} \setminus c_i) \cup \tilde{c}_i \cup c_{\mathrm{Elim},ij}. \tag{5.15}$$

Under the assumption that the only variable dropped from the constraint c_i by expressing it with the restriction $c_{\mathrm{Elim},ij}$ is z_j, the global structure graph of the constraint set in eqn. (5.15) is

$$G((\mathcal{C} \setminus c_i) \cup \tilde{c}_i \cup c_{\mathrm{Elim},ij}) = \{((\mathcal{C} \setminus c_i) \cup \tilde{c}_i \cup c_{\mathrm{Elim},ij}) \cup \mathcal{Z}, \mathcal{E}\}. \tag{5.16}$$

From the property given in eqn. (5.14) it is known that it is the edge e_{ij} which is locally ineffective. Thus, the graph in eqn. (5.16) can be directly obtained from the global structure graph of the constraint set $\mathcal{C} \cup c_{\mathrm{Elim},ij}$. This is done by eliminating the locally ineffective edge e_{ij} from the global structure graph $G(\mathcal{C} \cup c_{\mathrm{Elim},ij}) = \{\mathcal{C} \cup c_{\mathrm{Elim},ij} \cup \mathcal{Z}, \mathcal{E}_{\mathrm{global}}\}$:

$$G((\mathcal{C} \setminus c_i) \cup \tilde{c}_i \cup c_{\mathrm{Elim},ij}) = \{(\mathcal{C} \cup c_{\mathrm{Elim},ij}) \cup \mathcal{Z}, \mathcal{E}_{\mathrm{global}} \setminus e_{ij}\}. \tag{5.17}$$

If more than one edge is locally ineffective, the description of the operating region consists not only of one constraint $c_{\mathrm{Elim},ij}$ but of a set of constraints $\mathcal{C}_w$. Each of the elements of $\mathcal{C}_w$ corresponds to a locally ineffective edge in the set $\mathcal{E}_w$. Similar to the reasoning for one locally ineffective edge, for this more complex operating region $\mathcal{C}_w$, one obtains

$$(\mathcal{C} \setminus \mathcal{C}_i) \cup \tilde{\mathcal{C}}_i \cup \mathcal{C}_w \tag{5.18}$$

as an analytical model for the behavior of the plant in the operating region $\mathcal{C}_w$. The global structure graph of this constraint set is

$$G((\mathcal{C} \setminus \mathcal{C}_i) \cup \tilde{\mathcal{C}}_i \cup \mathcal{C}_w) = \{(\mathcal{C} \cup \mathcal{C}_w) \cup \mathcal{Z}, \mathcal{E}_{\mathrm{global}} \setminus \mathcal{E}_w\}. \tag{5.19}$$

This graph describes the couplings among the variables under the condition that the constraints in the set $\mathcal{C}_w$ are satisfied - e.g. the plant is in the operating region defined by $\mathcal{C}_w$. Using directed edges to represent causality one can therefore define:

Definition 5.3 (Directed local structure graph). *The directed structure graph of the constraint set $\mathcal{C}$, which is valid if the edges $\mathcal{E}_w$ are locally ineffective (that is the variables which appear in the constraint set $\mathcal{C}_w$ satisfy the constraint set $\mathcal{C}_w$) is* called *directed local structure graph in $\mathcal{C}_w$. It is denoted by*

$$G|\mathcal{C}_w = \{(\mathcal{C} \cup \mathcal{C}_w) \cup \mathcal{Z}, \mathcal{E} \setminus \mathcal{E}_w\}. \tag{5.20}$$

Loosely speaking, $G|\mathcal{C}_w$ means the structure graph which is valid in the operating region defined by $\mathcal{C}_w$. Note that the edges $\mathcal{E}$ are not only the edges of the directed global structure graph $G(\mathcal{C})$, but contain the edges between the constraint-vertices $\mathcal{C}_w$ and the variable-vertices as well.

Definition 5.3 reveals that the directed global structure graph is a special case of the directed local structure graph with the sets $\mathcal{C}_w = \emptyset$ and $\mathcal{E}_w = \emptyset$.

Similar to the usage on the global structure graph, the $\mathrm{var}^G(\cdot)$-operator can be applied to the local structure graph in order to determine the variables that have a direct impact on a constraint in a specific operating region.

The assumption that the only variable that disappears from the constraint c_i when introducing the restriction $c_{\mathrm{Elim},ij}$ is z_j is rather restrictive. Simplifying $\tilde{c}_i$ may lead to the vanishing of other variables than z_j from c_i as well. Also, there is no unique form of the constraint $\tilde{c}_i$. This is because solving $c_{\mathrm{Elim},ij}$ for one variable and injecting the result into c_i in order to make the variable z_j vanish from the constraint c_i may be done in different ways - depending on the number of variables in $c_{\mathrm{Elim},ij}$. Each of these ways results in a different form of $\tilde{c}_i$.

For these reasons, the actual set of variables that appears in a constraint $\tilde{c}_i$ may be a subset of the variable set

$$\mathrm{var}^{G|\mathcal{C}_w}(c_i) \tag{5.21}$$

which is determined using the local structure graph $G|\mathcal{C}_w$. This is an effect which cannot be taken into account by the local structure graph. However, the effect needs to be considered when using results obtained with local structural analysis.

Similar to the global incidence matrix $\boldsymbol{M}$ for the representation of the directed global structure graph G, the local incidence matrix $\boldsymbol{M}|\mathcal{C}_w$ is used to describe the directed local structure graph $G|\mathcal{C}_w$. A sketch of the local incidence matrix is depicted in Fig. 5.7.

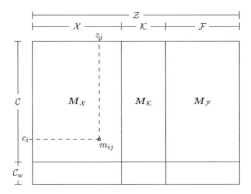

Figure 5.7: Sketch of the incidence matrix of a directed local structure graph

Due to the derivative in the operation to determine a constraint $c_{\mathrm{Elim},ij}$ under which the edge between the constraint-vertex c_i and the variable-vertex z_j is locally ineffective in Definition 5.2, only variables which appear in $c_{\mathrm{Elim},ij}$, may appear in c_i:

$$\mathcal{Z}_p \subseteq \mathcal{Z}_i. \tag{5.22}$$

Therefore, there are no additional variable-vertices in the directed local structure graph. This explains why the number of columns of the local incidence matrix is always the same as the number of columns in the global incidence matrix. The number of rows in the local incidence matrix is larger than the number of rows in the global incidence matrix. This is illustrated by the additional part at the bottom of the local incidence matrix in Fig. 5.7 in comparison to the global incidence matrix in Fig. 4.10.

An important observation is that the formalism $\mathrm{var}^{G|\mathcal{C}_w}(c_i)$ also allows to explain why an operating region in which a specific local structure graph $G|\mathcal{C}_w$ holds usually improves the structural discriminability: Assume that the edges $\mathcal{E}_w$ connect fault-vertices with the constraint c_i in the global structure graph G. Let the edges $\mathcal{E}_w$ be locally ineffective in $\mathcal{C}_w$. Then, these fault variables are not contained in

$$\mathcal{F} \cap \mathrm{var}^{G|\mathcal{C}_w}(c_i) \tag{5.23}$$

which is determined using the local structure graph in $\mathcal{C}_w$. This is the case, although they are contained in

$$\mathcal{F} \cap \mathrm{var}^{G}(c_i) \tag{5.24}$$

which is determined using the global structure graph. Then, these faults do not have an impact on c_i in the operating region $\mathcal{C}_w$ although they may influence the system-behavior in other operating regions. More generally,

$$\mathcal{F} \cap \mathrm{var}^{G|\mathcal{C}_w}(c_i) \subseteq \mathcal{F} \cap \mathrm{var}^{G}(c_i) \tag{5.25}$$

always holds. This means that by restricting the plant to a specific operating region $\mathcal{C}_w$ couplings between fault variables and the rest of the constraints in the analytical model may only vanish. This indicates that in the operating region $\mathcal{C}_w$, the impact of some faults is lost, whereas the impact of other faults isn't. This is generally beneficial to fault discrimination.

An example of a local structure graph $G|\mathcal{C}_w$ and the corresponding local incidence matrix $\boldsymbol{M}|\mathcal{C}_w$ is given in Example 5.3.

In this section, the consequence of a vanishing coupling for the structural representation of a constraint set $\mathcal{C}$ was given. Using the directed global structure graph, operating regions in which edges in this structure graph become locally ineffective can be identified. Because they represent a vanished coupling, the locally ineffective edges can be dropped from the global structure graph if it is ensured that the corresponding operating region is present. On the basis of this insight the directed local structure graph $G|\mathcal{C}_w$ was introduced to describe the couplings in a constraint set in a specific operating region. The directed local structure graph of the constraint set $\mathcal{C}$ in the operating region $\mathcal{C}_w$ is defined as the directed global structure graph of $\mathcal{C} \cup \mathcal{C}_w$ without the locally ineffective edges.

The subejct of the next section is the analysis of diagnosability properties which are proper to specific operating regions with the help of the directed local structure graph.

Example 5.3 (Directed local structure graph of Plant A). *According to the previous example, the edge e_2 is locally ineffective in $c_{\mathrm{Elim},2}$: $0 = x_3(t)$. The directed local structure graph $G|\{c_{\mathrm{Elim},2}\}$ is depicted in Fig. 5.8, its incidence matrix is given in Tab. 5.1. The edge e_2 is dashed to signify that it is locally ineffective. The corresponding entry $m_{5,1}$ in the local incidence matrix is zero.*

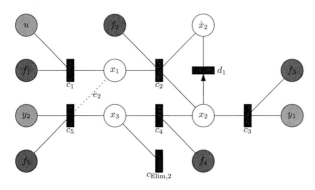

Figure 5.8: Directed local structure graph of Plant A

Table 5.1: Incidence matrix $M|\{c_{\mathrm{Elim},2}\}$ of the directed local structure graph $G|\{c_{\mathrm{Elim},2}\}$ of Plant A in $\{c_{\mathrm{Elim},2}\}$

	x_1	x_2	$\dot{x}_2$	x_3	u	y_1	y_2	f_1	f_2	f_3	f_4	f_5
m_1^T	1				1			1				
m_2^T	1	1	1						1			
m_3^T		1				1			1			
m_4^T		1		1						1		
m_5^T	0			1			1					1
$m_{d_1}^T$		-1	1									
m_w^T				1								

$$\underbrace{\qquad}_{M_{\mathcal{X}}|\{c_{\mathrm{Elim},2}\}} \quad \underbrace{\qquad}_{M_{\mathcal{K}}|\{c_{\mathrm{Elim},2}\}} \quad \underbrace{\qquad}_{M_{\mathcal{F}}|\{c_{\mathrm{Elim},2}\}}$$

5.5 Local Structural Properties of Constraint Sets

In this section, the link between the analysis of the directed local structure graph and the I/O-detectability and I/O-discriminability of faults is drawn. It is shown that the analysis of the local structure graph allows to infer on the I/O-diagnosability properties more precisely than the analysis of the global structure graph does. In that way, the restriction to specific operating regions allows more specific results of the structural analysis of an analytical model. Thus, the analysis of the directed local structure graph overcomes some of the limits of the global structural analysis.

The analysis of the directed global structure graph in Chapter 4 led to the formulation of "typically"-statements for the relationship between global structural properties and the I/O-detectability as well as global structural properties and the I/O-discriminability of faults. Also, it allowed to identify constraint sets which can be used to determine global residual generators which may indicate the presence of a fault. Since the global structure graph of an analytical model describes all couplings in a plant, independently of their local effectiveness, the results of the global structural analysis were conservative. In particular, as discussed in Section 4.9, the global structural analysis is limited in some situations: Counterexample 1 showed that with the global structural analysis one may find the result that a fault is not globally structurally detectable, although this fault actually is I/O-detectable in the sense of Definition 4.2. The reason for this is that the fault can only be detected if the plant has been brought to a specific operating region - a property which cannot be described by the global structure graph.

The directed local structure graph which was introduced in the previous section allows to describe structural peculiarities which are proper to an operating region. This motivates the idea to use this representation of the couplings in the plant to obtain more precise statements on the I/O-detectability and the I/O-discriminability of faults. Also, the results of the global structural analysis, the GMSOs could be used to solve the diagnostic problem via the design of global residual generators. Therefore, the idea to use the more precise results of the analysis of the directed local structure graph for the solution of the diagnostic problem is obvious. Two facts support the idea to investigate the local structure of a plant in order to analyze its I/O-diagnosability properties:

- Since the global structure graph contains an edge if it is locally effective in at least one operating point, it has the tendency to exaggerate the actual couplings in the analytical model. In particular, if the plant is restricted to an operating point or an operating region, couplings between variables and constraints may vanish, although the global structure graph indicates their presence for another operating point. Diagnosability properties which are proper to specific operating regions can therefore not be determined by investigating the global structure graph but rather the local structure graph. In particular, operating regions in which the coupling between a fault-variable-vertex and the rest of the graph vanishes are promising for fault discrimination.

- The local structure graph contains more constraint-vertices than the global structure graph, whereas the number of vertices representing the unknown variables is the same. This follows directly from the definition of local ineffectiveness in Definition 5.2:

$$\mathcal{Z}_p \subseteq \mathcal{Z}_i. \tag{5.26}$$

For that reason, the degree of structural redundancy $\bar{\varphi}$ of the local structure graph is generally larger than the degree of structural redundancy of the global structure graph. A larger degree of structural redundancy indicates better structural fault detectability and discriminability properties. Therefore, the analysis of the local structure graph may reveal detectability and discriminability which is proper to the operating region in which a specific local structure graph is valid.

Provided that the plant is in the operating region in which the local structure graph holds, the local structure graph describes the couplings in the analytical model correctly. Under this assumption on the actual operating point of the plant, the reasoning concerning the relationship between global structural properties in Chapter 4 and I/O-properties holds for the same analysis of the local structure graph as well. In particular, the definitions 4.3, 4.12 as well as the theorems 4.2, 4.3, 4.5, 4.7, 4.9 and 4.10 can be carried over and extended to the analysis of analytical models which are restricted to specific operating regions, the structure of which can be described by a local structure graph.

In the following, this reasoning is pursued for the relationship between the analysis of the directed local structure graph and the I/O-detectability of a fault.

For this purpose, the main statement of Theorem 4.2 is briefly recalled: A fault f_i is I/O-detectable if there is an input signal $\boldsymbol{u}(t)$ such that the constraint set $\mathcal{C}_{f_i}$ does not have a solution for the unknown variables in it, if this input signal $\boldsymbol{u}(t)$ and the fault-free output signal $\boldsymbol{y}_0(t)$ are injected into the constraint set: $\mathcal{C}_{f_i}\left((\boldsymbol{u}(t), \boldsymbol{y}_0(t))\right)$.

The assumption that the input signal $\boldsymbol{u}(t)$ steers the plant into the operating region $\mathcal{C}_w$ is simply an additional restriction on $\boldsymbol{u}(t)$ narrowing the space of possible input signals, c.f. the reasoning in Section 5.2. Provided that the plant is in an operating region defined by $\mathcal{C}_w$, the variables obey the constraint set $\mathcal{C} \cup \mathcal{C}_w$. Theorem 4.2 does not make any assumptions on the nature of the constraints which govern the plant behavior. Hence, if the plant is known to be in the operating region $\mathcal{C}_w$, a fault f_i is detectable, if the I/O-pair obtained from the fault-free plant does not allow the constraint set $\mathcal{C} \cup \mathcal{C}_w$ to have a solution for the unknown variables. It is therefore possible to formulate:

Lemma 5.1. *A fault f_i is I/O-detectable if there is an input signal $\boldsymbol{u}(t)$ which steers the plant into a given operating region $\mathcal{C}_w$ and the constraint set*

$$\mathcal{C}_{f_i,w}\left((\boldsymbol{u}(t), \boldsymbol{y}_0(t))\right) = \mathcal{C}_{f_i,w}\left((\boldsymbol{u}(t), \mathcal{C}_0 \circ \boldsymbol{u}(t))\right), \quad f_i \neq 0 \tag{5.27}$$

with $\mathcal{C}_{f_i,w} = \mathcal{C}_{f_i} \cup \mathcal{C}_w$ does not have a solution for the unknown variables $\mathcal{X}$, if $\boldsymbol{y}_0(t)$ is the output signal with which the fault-free plant answers to the excitation with the input signal $\boldsymbol{u}(t)$.

Proof. Let $\boldsymbol{u}(t)$ be an input signal which steers the plant into the operating region $\mathcal{C}_w$, independently of the fault-state of the plant. Then, the plant behavior is governed by the constraint set $\mathcal{C} \cup \mathcal{C}_w$, or more precisely in the case of a fault f_i by

$$\mathcal{C}_{f_i,w} = \mathcal{C}_{f_i} \cup \mathcal{C}_w. \tag{5.28}$$

Then, it is possible to interpret $\mathcal{C}_{f_i,w}$ as the analytical model of a new plant in which the fault f_i is present. Applying Theorem 4.2, one can infer on the I/O-detectability of fault f_i. □

Theorem 4.2 is a necessary and sufficient condition for the I/O-detectability of a fault, whereas Lemma 5.1 is a sufficient condition only. The reason for this major difference lies in the assumption of a given operating region: This assumption does not allow to make any statement on the I/O-detectability of a fault in operating regions other than $\mathcal{C}_w$.

Of course, $\mathcal{C}_{f_i,w}((\boldsymbol{u}(t), \boldsymbol{y}_0(t)))$ always having a solution for those input signals $\boldsymbol{y}_0(t)$ which steer the plant into the operating region $\mathcal{C}_w$ may also have its reason in a strong restriction of the input signals. This explains why it is not possible to infer from the non-existence of a $\boldsymbol{u}(t)$ which leads to a contradiction of $\mathcal{C}_{f_i,w}$ on the non-I/O-detectability of a fault for a given operating region $\mathcal{C}_w$. The extension of Lemma 5.1 from a given $\mathcal{C}_w$ to all possible operating regions, would allow this reasoning - however, investigating all possible operating regions separately is not possible. A solution to this problem would be to assume a $\mathcal{C}_w = \emptyset$ as the largest possible operating region. This corresponds to no restriction of the operating region and leads back to Theorem 4.2.

In order for a constraint set not to have a solution for the unknown variables in it, two properties must be fulfilled: The constraint set must be contradictable and the known variables must lead to this contradiction. Definition 4.3 for the first part of this reasoning allows to formulate:

Theorem 5.1. *Provided, there is an input signal $\boldsymbol{u}(t)$ which steers the plant into the operating region $\mathcal{C}_w$, a fault f_i is I/O-detectable if there is a subset*

$$\mathcal{C}_{\mathrm{U},f_i} \subseteq \mathcal{C}_{f_i} \cup \mathcal{C}_w \tag{5.29}$$

which is contradictable under excitation with this $\boldsymbol{u}(t)$ and $\boldsymbol{y}_0(t) = \mathcal{C}_0 \circ \boldsymbol{u}(t)$ leads to the contradiction of $\mathcal{C}_{\mathrm{U},f_i}$.

Proof. Let $\mathcal{C}_{\mathrm{U},f_i} \subseteq \mathcal{C}_{f_i} \cup \mathcal{C}_w$ be a contradictable constraint set under excitation with $\boldsymbol{u}(t)$. Furthermore, let $\boldsymbol{u}(t)$ steer the plant into the operating region $\mathcal{C}_w$. Then, there is a signal $\boldsymbol{y}(t)$ such that the constraint set

$$\mathcal{C}_{\mathrm{U},f_i}((\boldsymbol{u}(t), \boldsymbol{y}(t))) \tag{5.30}$$

has no solution for the unknown variables in it, c.f. Definition 4.3. If the fault-free output signal

$$\boldsymbol{y}(t) = \boldsymbol{y}_0(t) = \mathcal{C}_0 \circ \boldsymbol{u}(t) \tag{5.31}$$

leads to the contradiction of the constraint set $\mathcal{C}_{\mathrm{U},f_i}$, the I/O-pair $((\boldsymbol{u}(t), \boldsymbol{y}_0(t)))$ also leads to the contradiction of $\mathcal{C}_{f_i,w} = \mathcal{C}_{f_i} \cup \mathcal{C}_w$. This is the case because $\mathcal{C}_{\mathrm{U},f_i} \subseteq \mathcal{C}_{f_i} \cup \mathcal{C}_w$ holds and a superset $\mathcal{C}_{f_i} \cup \mathcal{C}_w$ of a set $\mathcal{C}_{\mathrm{U},f_i}$ which is already contradicted is contradicted, too.

The fact that the I/O-pair $(\boldsymbol{u}(t), \boldsymbol{y}_0(t))$ contradicts the constraint set $\mathcal{C}_{f_i,w}$, means that this constraint set does not have a solution for the unknown variables if the I/O-pair is injected in it. Then, Lemma 5.1 allows to conclude that the fault f_i is I/O-detectable.$\square$

The important improvement of Theorem 5.1 on Theorem 4.5 is the following: In order to show contradiction with the analytical model of the plant in the operating region $\mathcal{C}_w$, one needs to find a contradictable subset of $\mathcal{C}_{f_i} \cup \mathcal{C}_w$ (c.f. Theorem 5.1) instead of a contradictable subset of $\mathcal{C}_{f_i}$ (c.f. Theorem 4.5). Using a structural approach, it is generally easier to find a contradictable subset of $\mathcal{C}_{f_i} \cup \mathcal{C}_w$ than finding a contradictable subset of $\mathcal{C}_{f_i}$. The reason for this is that the degree of structural redundancy in a constraint set usually becomes larger if the number of unknown variables remains constant while the number of constraints is increased. Of course, this improvement requires the existence of an input signal which steers the plant into the operating region $\mathcal{C}_w$.

The contradictability of a constraint set $\mathcal{C}_{\mathrm{U},f_i} \subseteq \mathcal{C}_{f_i} \cup \mathcal{C}_w$ can be shown with a consistency relation which contains only known, that is input and output variables. For this purpose, the unknown variables which appear in the constraint set $\mathcal{C}_{\mathrm{U},f_i}$ are computed and injected into a remaining constraint. If it is then possible to choose an arbitrary input signal $\boldsymbol{u}(t)$ and an arbitrary signal $\boldsymbol{y}(t)$, such that the consistency relation is contradicted, the constraint set $\mathcal{C}_{\mathrm{U},f_i}$ is contradictable.

It is in this reasoning that the restriction to a specific operating region plays a decisive role: Let e_{ij} be an edge between the constraint-vertex c_i and the variable-vertex z_j in the directed global structure graph $G(\mathcal{C})$. Furthermore, let the edge e_{ij} be locally ineffective in the operating region $\mathcal{C}_w$, let the plant be in this operating region and let the constraint c_i be solved uniquely for another variable z_r than z_j. The value of the variable z_r can then be determined by a rearrangement of c_i using the values of the other variables in it - except the variable z_j, the value of which is not necessary. If the constraint c_i is contained in the set $\mathcal{C}_{\mathrm{U}}$ which is to be invesitigated for contradictability, the variable z_j is not used to determine the consistency relation. The contradictability of the constraint set $\mathcal{C}_{\mathrm{U}}$ therefore does not depend on the variable z_j. This effect is the impact of a specific operating region in which a coupling vanishes on the contradictability of a constraint set.

The information on actually present couplings is covered by the directed local structure graph $G|\mathcal{C}_w$. From the global structural analysis it is already known that - under an additional causality condition - proper structurally overconstrained constraint sets are typically contradictable. If obtained from the directed local structure graph in $\mathcal{C}_w$, such constraint sets are presumably contradictable if the plant is in the operating region $\mathcal{C}_w$. This reasoning motivates the following definition:

Definition 5.4 (Locally Proper Structurally Overconstrained in $\mathcal{C}_w$ (LPSO)). *A subset of the constraints $\mathcal{C}_{\mathrm{LPSO}} \subseteq \mathcal{C} \cup \mathcal{C}_w$ is called* locally proper structurally overconstrained *(LPSO) in $\mathcal{C}_w$, if there is a matching on the local structure graph $G|\mathcal{C}_w$ that is complete w.r.t. the unknown variables $\mathcal{X} \cap \mathrm{var}^{G|\mathcal{C}_w}(\mathcal{C}_{\mathrm{LPSO}})$ but not complete w.r.t. the constraints in $\mathcal{C}_{\mathrm{LPSO}}$.*

The following reasoning explains the relationship between the local proper structural overconstrainedness and the contradictability of a constraint set. It is based on the idea that a constraint set $\mathcal{C}_U$ is contradictable, if it is possible to compute the unknown variables in it from the known variables using at most

$$|\mathcal{C}_U \setminus c_k| = |\mathcal{C}_U| - 1 \qquad (5.32)$$

constraints of $\mathcal{C}_U$ and inject the result into the remaining constraint c_k:

1. Let $u(t)$ be an input signal that steers the plant into the operating region $\mathcal{C}_w$. Then the directed local structure graph $G|\mathcal{C}_w$ holds.

2. Let $\mathcal{C}_U \subseteq \mathcal{C} \cup \mathcal{C}_w$ be locally proper structurally overconstrained and let the input $u(t)$ be applied to the plant. Then there is a complete matching w.r.t. the unknown variables

$$\mathcal{X} \cap \mathrm{var}^{G|\mathcal{C}_w}(\mathcal{C}_U) \qquad (5.33)$$

which appear in $\mathcal{C}_U$ on the graph $G|\mathcal{C}_w$ and which have an impact on the manifolds defined by the constraints in the set $\mathcal{C}_U$. These are the variables which are connected to the constraints in the set $\mathcal{C}_U$ by edges which are locally effective in $\mathcal{C}_w$. It is possible that other variables occur in the constraints in the set $\mathcal{C}_U$. However, since these variables are connected to the constraints in the set $\mathcal{C}_U$ by locally ineffective edges in $\mathcal{C}_w$, they do not have an impact on the manifolds defined by the constraints in the set $\mathcal{C}_U$.

3. By removing one constraint c_k from $\mathcal{C}_U$, the property that there is a matching that is complete w.r.t. the unknown variables on the local structure graph $G|\mathcal{C}_w$ is preserved. One obtains the constraint set

$$\mathcal{C}_{LSJ} = \mathcal{C}_U \setminus c_k. \qquad (5.34)$$

The unknown variables which are connected to the constraints in the set $\mathcal{C}_{LSJ}$ by locally effective edges are the same as the unknown variables which are connected to the constraint in the set $\mathcal{C}_U$ by locally effective edges:

$$\mathcal{X} \cap \mathrm{var}^{G|\mathcal{C}_w}(\mathcal{C}_{LSJ}) = \mathcal{X} \cap \mathrm{var}^{G|\mathcal{C}_w}(\mathcal{C}_U). \qquad (5.35)$$

If there is a causal matching of the unknown variables in $\mathcal{X} \cap \mathrm{var}^{G|\mathcal{C}_w}(\mathcal{C}_{LSJ})$, it is typically possible to determine the values of the unknown variables in $\mathcal{C}_{LSJ}$ from the values of the known variables. For this purpose, it is not necessary to take into account the values of variables which are connected to the constraints in $\mathcal{C}_{LSJ}$ by locally ineffective edges. Computing the values of the unknown variables may require to use the information contained in $\mathcal{C}_w$.

4. Injecting the result for the unknown variables into the remaining constraint c_k, one obtains a consistency relation for which it is typically possible to choose a signal $y(t)$ that leads to contradiction of this relation.

The above reasoning allows to formulate the following property:

Remark 5.1. *A constraint set $\mathcal{C}_U \subseteq \mathcal{C} \cup \mathcal{C}_w$ is typically contradictable under excitation with the input signal $\mathbf{u}(t)$, if it is locally proper structurally overconstrained in $\mathcal{C}_w$, there is a causal matching on the directed local structure graph $G|\mathcal{C}_w$ which is complete w.r.t. the unknown variables and the input signal $\mathbf{u}(t)$ steers the plant into the operating region $\mathcal{C}_w$.*

Similar to the global structural analysis, the local structure graph allows to determine constraint sets $\mathcal{C}_U \subseteq \mathcal{C} \cup \mathcal{C}_w$ with the property that no proper subset of $\mathcal{C}_U$ is locally proper structurally overconstrained. In analogy to the global structural analysis, these constraint sets are called LMSO:

Definition 5.5 (Locally Minimal Structurally Overconstrained in $\mathcal{C}_w$ (LMSO)).
A subset of constraints $\mathcal{C}_{LMSO} \subseteq \mathcal{C} \cup \mathcal{C}_w$ is called locally minimal structurally overconstrained *(LMSO) in $\mathcal{C}_w$, if it is locally proper structurally overconstrained in $\mathcal{C}_w$, but no proper subset of $\mathcal{C}_{LMSO}$ is locally proper structurally overconstrained in $\mathcal{C}_w$.*

These smallest possible locally proper structurally overconstrained sets can be found by applying the algorithm FINDMSO to the local structure graph.

Example 5.4 (LMSOs of Plant A in $\mathcal{C}_2$). *The application of the algorithm FIND-MSO to the directed local structure graph $G|\mathcal{C}_2$ of Plant A (c.f. Example 5.3) which holds in the operating region $\mathcal{C}_2 = \{c_{Elim,2}\}$ with $c_{Elim,2}: \quad 0 = x_3(t)$ yields a total of six LMSOs. These LMSOs are:*

$$\mathcal{C}_{LMSO,2,1} = \{c_1, c_2, c_3, d_1\}$$
$$\mathcal{C}_{LMSO,2,2} = \{c_1, c_2, c_4, c_5, d_1\}$$
$$\mathcal{C}_{LMSO,2,3} = \{c_1, c_2, c_4, c_{Elim,2}, d_1\}$$
$$\mathcal{C}_{LMSO,2,4} = \{c_3, c_4, c_5\}$$
$$\mathcal{C}_{LMSO,2,5} = \{c_3, c_4, c_{Elim,2}\}$$
$$\mathcal{C}_{LMSO,2,6} = \{c_5, c_{Elim,2}\}.$$

Note that these are more than the GMSOs obtained with the analysis of the global structure graph, c.f. Example 4.5.

Obviously, the analysis of the local structure graph does not only result in more LMSOs than the analysis of the global structure graph results in GMSOs but also in different constraint sets.

Although the global structural detectability of a fault f_i is neither a necessary nor a sufficient condition for the I/O-detectability of f_i, c.f. Section 4.9, this analysis gives a hint on the I/O-detectability of a fault. Similarly, the results from the local structural analysis indicate I/O-detectability: The set of LMSOs allows to conclude which faults are structurally detectable under the condition that the plant has been brought to the operating region $\mathcal{C}_w$. Based on the definition of LMSOs, the idea of local structural detectability of a fault is introduced in the following. In analogy to the global structural detectability of a fault, the corresponding local property of a fault f_i is defined as:

Definition 5.6 (Local structural detectability). *A fault is called locally structurally detectable in $\mathcal{C}_w$, if*

- *there is a locally minimal structurally overconstrained constraint set $\mathcal{C}_{\mathrm{LMSO}}$ such that*

$$f_i \in \mathrm{var}^{G|\mathcal{C}_w}(\mathcal{C}_{\mathrm{LMSO}}) \qquad (5.36)$$

holds,

- *there is a causal matching that is complete w.r.t. the unknown variables in $\mathcal{X} \cap \mathrm{var}^{G|\mathcal{C}_w}(\mathcal{C}_{\mathrm{LMSO}})$ on the local structure graph $G|\mathcal{C}_w$, and*

- *there is an input signal $\boldsymbol{u}(t)$ that steers the plant into the operating region $\mathcal{C}_w$.*

In the following, no relationship between LMSOs and the I/O-discriminability of two faults will be established. The transfer of the reasoning concerning the relationship between local structural detectability, LMSOs, contradictability and I/O-detectability described in this section can be carried over to the local structural discriminability. Although this is not detailed here, similar to the carry-over from global structural detectability to local structural detectability, local structural discriminability as a counterpart to global structural discriminability is defined in the following:

Definition 5.7 (Local structural discriminability). *A fault f_i is called locally structurally discriminable from fault f_j in the operating region $\mathcal{C}_w$, if*

- *there is a locally minimal structurally overconstrained constraint set $\mathcal{C}_{\mathrm{LMSO}}$ such that*

$$f_i \in \mathrm{var}^{G|\mathcal{C}_w}(\mathcal{C}_{\mathrm{LMSO}}) \wedge f_j \notin \mathrm{var}^{G|\mathcal{C}_w}(\mathcal{C}_{\mathrm{LMSO}}) \qquad (5.37)$$

holds,

- *there is a causal matching that is complete w.r.t. the unknown variables in $\mathcal{X} \cap \mathrm{var}^{G|\mathcal{C}_w}(\mathcal{C}_{\mathrm{LMSO}})$ on the local structure graph $G|\mathcal{C}_w$, and*

- *there is an input signal $\boldsymbol{u}(t)$ that steers the plant into the operating region $\mathcal{C}_w$.*

Since the LMSOs differ from the GMSOs, it is self-evident that - provided the plant is in the operating region in which the local structure graph $G|\mathcal{C}_w$ holds - the local structural diagnosability properties differ from the global ones. If all LMSOs are determined, similar to the global signature matrix, the local signature matrix $\boldsymbol{S}|\mathcal{C}_w$ in the operating region $\mathcal{C}_w$ can be used to investigate local structural detectability and discriminability.

Example 5.5 (Local structural detectability of faults in Plant A). *In this example, the local structural detectability of the faults in Plant A in the operating region*

$$\mathcal{C}_2 = \{c_{\text{Elim},2}\} \quad with \quad c_{\text{Elim},2}: \quad 0 = x_3(t) \tag{5.38}$$

is investigated. One condition in the definition of local structural detectability and discriminability is the existence of an input signal which steers the plant into an operating region in which the local structure graph holds. This is given here, since choosing an input signal $u(t)$ which leads to $y_1(t) = 0$ and, therefore, $x_2(t) = x_3(t) = 0$ is possible. Of course, this is more difficult in the faulty case than in the nominal case. This problem will be addressed later. For all LMSOs known from Example 5.4, there is a causal matching that is complete w.r.t. the unknown variables which appear in the LMSO on the local structure graph $G|\mathcal{C}_w$. The local signature matrix $\boldsymbol{S}|\mathcal{C}_w$ is given in Tab. 5.2

Table 5.2: Plant A: Local signature matrix in $\mathcal{C}_2$

$\boldsymbol{S}|\mathcal{C}_w$	f_1	f_2	f_3	f_4	f_5
$\mathcal{C}_{\text{LMSO},1}$	1	1	1	0	0
$\mathcal{C}_{\text{LMSO},2}$	1	1	0	1	1
$\mathcal{C}_{\text{LMSO},3}$	1	1	0	1	0
$\mathcal{C}_{\text{LMSO},4}$	0	0	1	1	1
$\mathcal{C}_{\text{LMSO},5}$	0	0	1	1	0
$\mathcal{C}_{\text{LMSO},6}$	0	0	0	0	1

From the local signature matrix $\boldsymbol{S}|\mathcal{C}_w$, it is possible to conclude that all faults are locally structurally detectable in $\mathcal{C}_w$ since there is an entry for each fault in a row. Also, all faults except f_1 and f_2 are locally structurally discriminable in $\mathcal{C}_w$. This is a significant difference w.r.t. the results of the global structural analysis, c.f. Example 4.8. Whereas the faults f_1 and f_2 are globally structurally discriminable, but the faults f_4 and f_5 are not, the local structural analysis results in the opposite. Obviously, the plant's restriction to specific operating regions results in specific structural diagnosability properties. Note that here the difference between the local signature matrix and the global signature matrix was not a result from a vanished coupling between a fault-variable-vertex and a constraint-vertex. Instead, a coupling between an unknown variable and a constraint vanished.

The relationship between the local structural detectability and the I/O-detectability is elaborated in the following. From Section 4.8 we know that although a GMSO cannot be guaranteed to be contradictable, this is often the case. This motivates the analysis of the relationship between local structural properties - in particular LMSOs which are the basis of local structural detectability - and the contradictability of constraint sets.

The following reasoning holds: Let the fault f_i be locally structurally detectable in $\mathcal{C}_w$. Then, the fault variable f_i appears in a locally minimal structurally overconstrained constraint set, c.f. Definition 5.6. Typically, under the assumption that the input signal $\boldsymbol{u}(t)$ steers the plant into the operating region $\mathcal{C}_w$, the LMSO is contradictable under excitation with $\boldsymbol{u}(t)$. If the output signal that results from the fault-free plant contradicts this contradictable constraint set, the fault is I/O-detectable, c.f. Theorem 5.1. The following remark summarizes this reasoning.

Remark 5.2 (Local structural detectability and I/O-detectability). *A fault f_i is typically I/O-detectable if*

- *there is an operating region $\mathcal{C}_w$ in which f_i is locally structurally detectable,*

- *there is an input signal $\boldsymbol{u}(t)$ that steers the plant into the operating region $\mathcal{C}_w$ and*

- $\boldsymbol{y}_0 = \mathcal{C}_0 \circ \boldsymbol{u}(t)$ *contradicts* $\mathcal{C}_{\mathrm{LMSO}}$ *with* $f_i \in \mathrm{var}^{G|\mathcal{C}_w}(\mathcal{C}_{\mathrm{LMSO}})$.

One main condition for the I/O-detectability of a fault is the existence of a constraint set which is contradictable. In Remark 5.2, this is covered by an LMSO. It is here where a first major difference between the global and local structural analysis can be found: Whereas

- globally minimal structurally overconstrained constraint sets are typically contradictable under *arbitrary* excitation (the property does not depend upon the input signal $\boldsymbol{u}(t)$,

- locally minimal structurally overconstrained constraint sets are typically contradictable only under a specific excitation $\boldsymbol{u}(t)$: namely the one which steers the plant into the operating region in which the local structure graph used to obtain the LMSO holds.

Of course, local structural analysis suffers from a couple of limits which are very similar to the ones of global structural analysis. The reason for this is that neither global nor local structural analysis is able to take into account the influence of parameters (c.f. counterexample 3 in Section 4.9). This is why local structural analysis results in "typically"-statements like global structural analysis does. However, one significant difference with respect to global structural analysis can be pointed out: whereas global structural analysis is not able to exploit structural peculiarities in specific operating regions, local structural analysis does. It therefore resolves the limit presented in counterexample 1 in Section 4.9. The result of the global structural analysis of this example was that the fault f_1 is not globally structurally detectable although it is in fact I/O-detectable. Analyzing this example in the local structure graph in $\mathcal{C}_w = \{c_{\mathrm{Elim}} : \ 0 = u(t)\}$ has the result that the fault is locally structurally detectable in $\mathcal{C}_w$. Because the plant can be brought into the operating region $\mathcal{C}_w$ by applying $u(t) = 0$ to its input and because the output signal $y_0(t) = \mathcal{C}_0 \circ u(t)$ contradicts $\mathcal{C}_{f_i} \cup \mathcal{C}_w$, the fault is I/O-detectable. This example shows that the property local structural detectability indicates I/O-detectability more precisely than global structural detectability does.

The assumption that the plant has been brought to the operating region $\mathcal{C}_w$, allows to look for a contradictable subset of $\mathcal{C}_{f_i} \cup \mathcal{C}_w$ instead of a contradictable subset of $\mathcal{C}_{f_i}$ when analyzing the I/O-detectability of the fault f_i. This does in fact not improve the I/O-detectability of the fault f_i (which is proper to the plant and the fault and can hence not be changed by restricting the plant to a specific operating region). However, it may improve the reasoning on the I/O-detectability. This is the case because it is easier to find a contradictable subset in $\mathcal{C}_{f_i} \cup \mathcal{C}_w$ than in $\mathcal{C}_{f_i}$. This is the major difference between the global structural analysis and the local structural analysis.

The approach in [81] is equivalent to the analysis of the global structural detectability and isolability properties of the different analytical models that hold in the different discrete modes of a hybrid system. This approach can be interpreted as the investigation of local diagnosability properties. It is similar to the approach in this thesis in the way that it considers only specific operating regions. However, for two reasons the approach in [81] is limited in comparison to the approach taken here. First, in [81] the discrete modes are known prior to the analysis. Second, the description of the operating region is already incorporated in the model which is valid in a specific discrete mode.

Another major difference between the local structural analysis presented in this section and the one in [81] is that instead of predefined discrete modes of the plant, an analysis of the analytical model resulted in the description of an operating region. Also, in the approach presented here, no new analytical model was formulated for each operating region and analyzed structurally in a second step. Instead, the local structure graph resulted directly from the global structure graph with modifications taking the operating region into account.

Similar to the assumption that the input signal $\boldsymbol{u}(t)$ applied to the plant successfully steers the plant into the operating region $\mathcal{C}_w$, in [81] it was assumed that steering in the discrete mode in which the analysis holds is successful.

In this section it was shown that the restriction of a plant to specific operating regions allows specific and more precise results of the structural analysis of the diagnosability of faults. For this purpose, the local structural properties of a constraint set and their relationship to the I/O-detectability were investigated. Starting from the definition of I/O-detectability, a relationship between the existence of a solution for the unknown variables in an analytical model of plants which are steered into a specific operating region was established in Lemma 5.1. An intermediate result in Theorem 5.1 was that a fault is I/O-detectable if two conditions hold: First, there needs to be an input signal which steers the plant into a specific operating region. Second, the observed I/O-pair has to lead to contradiction with subsets of the constraint set describing the plant and the operating region. Remark 5.1 stated that locally proper structurally overconstrained constraint sets are typically contradictable under the excitation which steers the plant into a specific operating region. This is the operating region in which the local structure graph that was used to determine the LPSO holds. The application of the algorithm FINDMSO to a local structure graph results in locally minimal structurally overconstrained constraint sets which led - analogous to the global structural detectability - to the local structural detectability of a fault. Remark 5.2 described the relationship between local structural detectability and I/O-detectability. It revealed that it is not necessary that a fault is globally structurally detectable in order to be able to detect it.

However, since the local structure graph only holds in a specific operating region, properties like local structural detectability only hold if steering the plant into the operating region is successful. Two questions arise from the results in this section:

1. How can one exploit an LMSO for fault detection?

2. How can one steer a potentially faulty system into an operating region, such that the assumption that the plant has been brought to the operating region in which a constraint set is LMSOs holds?

The first question is addressed in the next section. The second question is dealt with in Section 6.4 and Section 6.6.

5.6   Local Residual Generator Design

In this section, the definition of local residuals is given and a method is presented which allows to determine local residual generators which provide local residuals.

For a given plant model and actuator/sensor configuration, the methods described in Section 4.2 generally result in very few global residual generators. A Boolean logic which exploits the information of nonzero global residuals is therefore very conservative with respect to fault discrimination. This major drawback allows to conclude that the use of a bank of global residual generators and a Boolean logic is inappropriate for service diagnosis.

However, if the plant is known to be in the operating region $\mathcal{C}_w$, it is possible to check the consistency of the present I/O-pair with the nominal behavior of the plant in the operating region $\mathcal{C}_w$ only. This local consistency test can be accomplished by a new kind of signals which only has the properties of residuals if the plant is in the operating region $\mathcal{C}_w$. Thus, the rather restrictive condition for global residuals to be zero for *all* input signals and output signals of the plant in the nominal case is relaxed. This leads to the concept of *local residuals* which can be used in automatic tests, c.f. Chapter 6:

Definition 5.8 (Local residual).   *The signal $r_w(t)$ is called* local residual in $\mathcal{C}_w$, *if the following properties hold:*

- *If the plant is in the operating region $\mathcal{C}_w$ and there is no fault in the plant, $r_w(t) = 0 \; \forall \; t$ holds.*

- *If the plant is in the operating region $\mathcal{C}_w$ and $r_w(t) \neq 0$ holds, the plant is faulty.*

A dynamical system which computes a local residual $r_w(t)$ from the input signals and the output signals of the plant is called local residual generator in $\mathcal{C}_w$. *It is denoted by*

$$r_w(t) = r_w\left(\boldsymbol{u}(t), \boldsymbol{y}(t), \frac{\mathrm{d}}{\mathrm{d}t}\right). \tag{5.39}$$

Again, the notation $\frac{d}{dt}$ is used to denote that not only the signals $\boldsymbol{u}(t)$ and $\boldsymbol{y}(t)$ but also their derivatives w.r.t. time of any order may occur in the local residual generator.

Global residual generators (c.f. Definition 4.1) check whether an observed I/O-pair is consistent with the entire nominal behavior $\mathcal{B}_0$. In contrast to this, local residual generators check the consistency of an I/O-pair with a part of the nominal behavior only: A local residual $r_w(t)$ in $\mathcal{C}_w$ may take on arbitrary values if the plant is not in the operating region defined by $\mathcal{C}_w$. Therefore, if the plant is not in the operating region $\mathcal{C}_w$, the local residual in $\mathcal{C}_w$ does not allow to infer on the fault-state of the plant.

According to Section 5.2, an operating region can be mapped to a restriction of the I/O-pairs. This allows to conclude that the local residual relaxes a restrictive property of the global residual: Whereas the global residual is zero in the fault-free case for *arbitrary* input signals, the local residual is zero in that case for *specific* signals only – namely those which steer the plant into the operating region $\mathcal{C}_w$. In the illustration of such a restriction of the I/O-pairs in Fig. 5.2 and Fig. 5.3, a local residual in $\mathcal{C}_w$ only allows to infer on the fault-state of the plant in the shaded areas of the I/O-plane. A more detailed description of this is given with the help of Fig. 5.9 later on.

Global residual generators can be found on the basis of subsets of the constraint set which describes the nominal behavior in an arbitrary operating point. These subsets are determined by the analysis of the global structure graph.

In contrast, local residual generators check consistency of an observed I/O-pair with the nominal behavior in a specific operating region. They are therefore determined from subsets of the constraint set which describes the behavior of the nominal plant in the operating region: the union of $\mathcal{C}_0$ and $\mathcal{C}_w$. The couplings in the plant, which is restricted to a specific operating region, are described by the directed local structure graph. This graph therefore allows to determine the constraint subsets for determining local residual generators.

The above reasoning for the design of local residual generators is detailed in the following: Let $\mathcal{C}_0$ be the constraint set which describes the nominal behavior of the plant. If the plant is in the operating region $\mathcal{C}_w$, the variables satisfy not only $\mathcal{C}_0$ but also

$$\mathcal{C}_0 \cup \mathcal{C}_w. \tag{5.40}$$

Moreover, the variables satisfy any rearrangement $\tilde{\mathcal{C}}_0 \cup \mathcal{C}_w$ of this set, the structure of which is described by the local structure graph $G|\mathcal{C}_w$. Checking the consistency of an I/O-pair with a set of constraints requires this set of constraints to be contradictable. From the previous section it is already known that LMSOs in $\mathcal{C}_w$ are typically contradictable under the excitation with the input signal which steers the plant into the operating region $\mathcal{C}_w$ – provided there is a causal matching that is complete w.r.t. the unknown variables. Therefore, LMSOs are the basis for the design of local residual generators.

Let $\mathcal{C}_{\mathrm{LMSO}}$ be locally minimal structurally overconstrained in $\mathcal{C}_w$ with a causal matching that is complete w.r.t. the unknown variables on the local structure graph $G|\mathcal{C}_w$. It is then possible to determine the value of all unknown variables in the set

$$\mathcal{X} \cap \mathrm{var}^{G|\mathcal{C}_w}(\mathcal{C}_{\mathrm{LMSO}}) \tag{5.41}$$

using only the constraints in a subset $\mathcal{C}_{\mathrm{LMSO}} \setminus c$ of the LMSO. If the result is injected into the remaining constraint c and all fault variables are set to zero, one obtains a relation that contains only inputs and outputs. Specific I/O-pairs are consistent with this relation. These are the I/O-pairs which result from the excitation of the fault-free plant with the input signals that steer the plant into the operating region $\mathcal{C}_w$. If all terms of this relation are brought to one side, the other side forms a signal with the following properties:

1. The signal is zero if the input and output signals of the fault-free plant are injected and the plant is in the operating region defined by $\mathcal{C}_w$.

2. If the signal is nonzero and the plant is in the operating region defined by $\mathcal{C}_w$, a fault which has an impact on $\mathcal{C} \cup \mathcal{C}_w$, that is a fault in the set

$$\mathcal{F} \cap \text{var}^{G|\mathcal{C}_w}(\mathcal{C}_{\text{LMSO}}) \qquad (5.42)$$

is known to be present.

3. If the signal is nonzero and the plant is not in the operating region $\mathcal{C}_w$, no conclusion on the fault-state of the plant is possible.

The signal obtained in the described way therefore forms a local residual in the sense of Definition 5.8.

Because several LMSOs can be found by analyzing the local structure graph $G|\mathcal{C}_w$, they are denoted with the additional subscript s: $\mathcal{C}_{\text{LMSO},w,s}$. From each LMSO one can obtain a local residual. It is therefore denoted by $r_{w,s}(t)$.

If a local residual is obtained from the constraint set $\mathcal{C}_{\text{LMSO},w,s}$ and the plant is in the operating region $\mathcal{C}_w$, only the faults which occur in the constraints in $\mathcal{C}_{\text{LMSO},w,s}$ may lead to the local residual being nonzero. Different fault variables occur in the different constraint sets $\mathcal{C}_{\text{LMSO},w,s}$, thus influencing the sensitivity of the consistency check. This is one degree of freedom which will be used in the design of automatic tests in Chapter 6.

The procedure to determine local residuals is summed up in the following algorithm:

Algorithm 3 [FindLocalResiduals]:

Input: A constraint set $\mathcal{C}$ from an analytical model and an operating region $\mathcal{C}_w$.

Initialisation: An empty set $\mathcal{A}_{\text{Res},w}$.

Step 1: Determine the local structure graph $G|\mathcal{C}_w$.

Step 2: Use $G|\mathcal{C}_w$ and FINDMSO to determine all LMSOs $\mathcal{C}_{\text{LMSO},w,s} \subseteq \mathcal{C} \cup \mathcal{C}_w$.

Step 3: If there is a causal matching which is complete w.r.t. the unknown variables in $\mathcal{C}_{\text{LMSO},w,s}$ on the directed local structure graph $G|\mathcal{C}_w$, add $\mathcal{C}_{\text{Res},w,s} = \mathcal{C}_{\text{LMSO},w,s}$ to $\mathcal{A}_{\text{Res},w}$.

Result: A family $\mathcal{A}_{\text{Res},w}$ of sets, each element $\mathcal{C}_{\text{Res},w,s}$ of which allows to determine a local residual generator in $\mathcal{C}_w$ by elimination of the unknowns in $\mathcal{C}_{\text{Res},w,s}$.

In the following, the Algorithm FINDLOCALRESIDUALS is applied to the running example.

Example 5.6 (Local residuals for Plant A). *In order to illustrate the above algorithm, Plant A (c.f. Section 3.4) and the operating region $\mathcal{C}_2 = \{c_{\text{Elim},2}\}$ with $c_{\text{Elim},2}$: $0 = x_3(t)$ (c.f. Example 5.1 and Example 5.2) are considered.*

1. *The local structure $G|\mathcal{C}_2$ of the plant is found, c.f. Example 5.3.*

2. *The structural analysis of $G|\mathcal{C}_2$ yields a total of six LMSOs in $\mathcal{C}_2$. This result is already known from Example 5.4:*

$$\mathcal{C}_{\text{LMSO},2,1} = \{c_1, c_2, c_3, d_1\}$$
$$\mathcal{C}_{\text{LMSO},2,2} = \{c_1, c_2, c_4, c_5, d_1\}$$
$$\mathcal{C}_{\text{LMSO},2,3} = \{c_1, c_2, c_4, c_{\text{Elim},2}, d_1\}$$
$$\mathcal{C}_{\text{LMSO},2,4} = \{c_3, c_4, c_5\}$$
$$\mathcal{C}_{\text{LMSO},2,5} = \{c_3, c_4, c_{\text{Elim},2}\}$$
$$\mathcal{C}_{\text{LMSO},2,6} = \{c_5, c_{\text{Elim},2}\}$$

3. *There is a causal matching which is complete w.r.t. the unknown variables in the LMSOs on the graph $G|\mathcal{C}_2$ found in Step 2. Therefore, all the LMSOs $\mathcal{C}_{\text{LMSO},w,s}$ are called $\mathcal{C}_{\text{Res},w,s}$ and added to the familiy of sets $\mathcal{A}_{\text{Res},w}$.*

The result of FINDLOCALRESIDUALS *is*

$$\mathcal{A}_{\text{Res},w} = \{\mathcal{C}_{\text{Res},2,1}, \mathcal{C}_{\text{Res},2,2}, \mathcal{C}_{\text{Res},2,3}, \mathcal{C}_{\text{Res},2,4}, \mathcal{C}_{\text{Res},2,5}, \mathcal{C}_{\text{Res},2,6}\}. \tag{5.43}$$

By setting the fault variables to zero and eliminating the unknown variables from the constraint sets $\mathcal{C}_{\text{Res},w,s}$, one obtains the local residuals

$$r_{2,1}(t) = K_u K_{y1} u(t) - \frac{1}{T} y_1(t) - \frac{d}{dt} y_1(t)$$
$$r_{2,2}(t) = y_2(t)$$
$$r_{2,3}(t) = K_M u(t)$$
$$r_{2,4}(t) = y_2(t)$$
$$r_{2,5}(t) = \frac{K_M}{K_{y1}} y_1(t)$$
$$r_{2,6}(t) = y_2(t).$$

Note that these are different signals than the global residuals, c.f. Example 4.7.

Figure 5.9 illustrates the properties of a local residual generator in the I/O-plane. Assume that a local residual in $\mathcal{C}_w$ was obtained from a locally minimal structurally overconstrained constraint set $\mathcal{C}_{\text{LMSO}}$ which was found by analyzing the directed local structure graph $G|\mathcal{C}_w$. The set $\mathcal{B}_0$ is the behavior of the plant in the fault-free case, the set $\mathcal{B}(\mathcal{C}_{\text{LMSO}})$ corresponds to the I/O-pairs which are consistent with $\mathcal{C}_{\text{LMSO}}$ and, therefore, lead to the local residual being zero. The area which is shaded in light gray represents the set of I/O-pairs which correspond to the operating region described by $\mathcal{C}_w$. As for a global residual, the I/O-pair $(\boldsymbol{u}_A(t), \boldsymbol{y}_A(t))$ is inconsistent with the behavior $\mathcal{B}(\mathcal{C}_{\text{LMSO}})$ and the

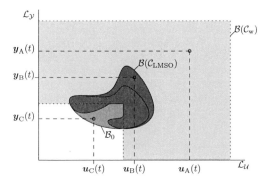

Figure 5.9: Consistency test realized by a local residual generator in the I/O-plane

behavior $\mathcal{B}_0$. Therefore, it indicates the presence of a fault. The I/O-pair $(\boldsymbol{u}_\mathrm{B}(t), \boldsymbol{y}_\mathrm{B}(t))$ is consistent with the behavior $\mathcal{B}(\mathcal{C}_\mathrm{LMSO})$. Hence, the corresponding local residual is zero which does not allow any conclusion on the fault-state of the plant. The difference of the local residual generator with respect to a global residual generator is that the I/O-pair $(\boldsymbol{u}_\mathrm{C}(t), \boldsymbol{y}_\mathrm{C}(t))$ is not consistent with the behavior $\mathcal{B}(\mathcal{C}_\mathrm{LMSO})$ and, therefore, the local residual is nonzero. This is the case although $(\boldsymbol{u}_\mathrm{C}(t), \boldsymbol{y}_\mathrm{C}(t))$ is consistent with the nominal behavior $\mathcal{B}_0$. However, since the I/O-pair $(\boldsymbol{u}_\mathrm{C}(t), \boldsymbol{y}_\mathrm{C}(t))$ is not in the behavior defined by $\mathcal{C}_w$, the fact that the local residual is nonzero does not allow any conclusion on the fault-state of the plant.

The example at hand also allows to formulate a general statement on the behavior defined by GMSOs and LMSOs. Whereas for globally minimal structurally overconstrained constraint sets

$$\mathcal{B}(\mathcal{C}_\mathrm{GMSO}) \subseteq \mathcal{B}_0 \qquad\qquad (5.44)$$

always holds, this is not the case for locally minimal structurally overconstrained constraint sets:

$$\mathcal{B}(\mathcal{C}_\mathrm{LMSO}) \nsubseteq \mathcal{B}_0. \qquad\qquad (5.45)$$

Note that there is a fundamental difference between the principle which is used to discriminate faults with the help of a bank of residual generators and the discrimination reached with the help of local residuals. The bank-of-residual-generators-approach tries to find as many global residuals as possible. In that way, it partitions the set of all I/O-pairs into sets, all of which contain the entire nominal behavior $\mathcal{B}_0$. This is explained by eqn. (5.44) and illustrated in Fig. 4.5. The approach is limited by the number of possible global residual generators, that is the number of GMSOs.

The local-residual-generator-approach partitions the I/O-plane into a part in which specific signals have the properties of a residual and into a part in which these signals do not have the properties of a residual. This is depicted in Fig. 5.9. Therefore, the behavior with which a local residual checks consistency does not necessarily contain the entire nominal behavior which is the meaning of eqn. (5.45).

It is this relaxation of the rather restrictive properties of global residuals which typically allows to find more local residuals than global residuals. In that way, more consistency checks can be obtained and faults which cannot be distinguished using global residuals may be discriminated, provided that the system has actually been brought to the operating region described by $\mathcal{C}_w$. The problem of steering a potentially faulty plant into a specific operating region is addressed in Section 6.4.

This section introduced the local residual, a signal which has the properties of a residual, provided the plant has been brought to a specific operating region. A method to determine local residual generators was proposed. Local residuals are an important part of automatic tests, the design of which is described in Chapter 6.

5.7 Summary

In this chapter, the idea that specific operating regions lead to a structural breakdown in the global structure graph of the plant was used to develop a new kind of signal, the local residual. This signal allows a consistency test of the observed I/O-pair presuming that the plant has been brought to the specific operating region.

The train of thought from the observation of the structural breakdown in specific operating regions to the local residual was as follows:

The global structure graph exaggerates the couplings in the plant. This is the case because it contains edges between variable-vertices and constraint-vertices although a variable may not have any impact on a constraint in a specific operating region. If the plant is restricted to such an operating region, the edge between this variable-vertex and this constraint-vertex can be dropped from the global structure graph. Removing the edge and adding constraints which describe the corresopnding operating region results in the local structure graph. This graph describes the couplings in the plant if the assumption that the plant is in the specific operating region holds. The consequences of the existence of such a graph for the analysis of the diagnosability properties of faults were analyzed. A method to determine specific constraint sets, the LMSOs which are typically contradictable under a specific excitation, was presented. This method makes use of the local structure graph and the algorithms which are already known from the analysis of the global structure graph. Eliminating the unknown variables from LMSOs allows to find local residual generators.

The benefits of local structural analysis are a more precise analysis of the I/O-diagnosability of faults and a method to determine consistency checks with better discrimination properties than the consistency checks found with global structural analysis. The price for this improvement is the difficult task of steering a potentially faulty system into specific operating regions.

The local residuals and their properties play an important role in the next chapter, in which a design method for automatic tests is presented.

Chapter 6

Design of Diagnostic Tests

6.1 Extension to a Method for the Design of Automatic Tests

This chapter is concerned with merging the previously elaborated results to a novel method for the design of automatic tests in order to solve the problem stated in Section 2.7. An automatic test consists of two parts: the test signal generator and the diagnostic unit. A block diagram of the latter is depicted in Fig. 6.1.

The idea behind the diagnostic unit is to use a local residual generator to perform the consistency test once it is detected that a specific operating region is present. A decision logic then decides which fault-hypotheses can be rejected and which hypotheses need to be preserved. A procedure to determine local residuals was already described in the previous chapter. The first two main contributions of this chapter therefore concern a design method for the decision logic and a method for testing whether the plant is in the desired operating region.

The detection of the operating region is achieved by a new kind of signals called validuals. A validual indicates the presence of an operating region in which an edge in the global structure graph is locally ineffective. In that way, a number of validuals can be used to infer on an operating region in which a specific local structure graph, namely the one used to determine the local residual, is present. A procedure to determine validuals is given in the algorithm FINDVALIDUALS. It makes use of the directed global structure graph.

The decision logic is designed with the help of the structure graphs which are used to determine the local residual and the validuals. It is based on Theorem 6.1 which states that the validuals indicating an operating region and the local residual being nonzero may have its reason in both, a failed detection of the operating region and an inconsistency detected by the local residual.

In order to account for this, the operating region detection is designed for the fault-free plant. Then the decision logic must take into account the faults which may lead to an erroneous detection of the operating region. The important result of this design procedure is that the diagnostic unit does not reject true fault-hypotheses, although steering in the desired operating region may have failed. In that way the difficult problem of steering a potentially faulty system is reduced to the detection of an operating region in the nominal system and the appropriate design of the decision logic.

The third important contribution concerns the test signal generator. Since the validuals being zero indicates that the operating region was reached, this input generator is designed in a way to render the validuals zero. Because the expression of the validual is known, it can be solved for the inputs of the system, thus forming the input generator. An interesting observation is that depending on the validual, the input generator may consist of a feedforward controller, a feedback controller or a control law.

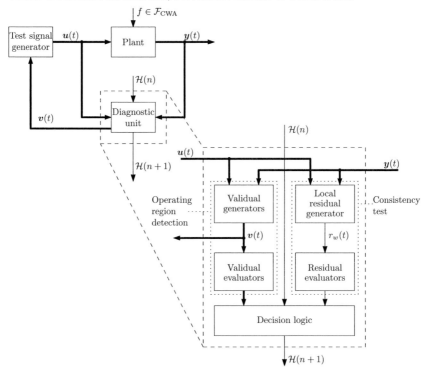

Figure 6.1: Automatic test and its diagnostic unit

The chapter is organized as follows: The interaction of the components of an automatic test is described in Section 6.2. The diagnostic unit is treated in Section 6.3. The method for the design of the validuals which are used to infer on the presence of an operating region can be found in Section 6.4. A graph-based method to determine which fault-hypotheses can be rejected with a test is developed in Section 6.5. Section 6.6 introduces the method to determine the test signals and in Section 6.7 an algorithm to compute all possible tests is given. This algorithm is based on the analysis of global and local structure graphs. In Section 6.8 a special case of the tests, the fault hiding, is explained.

6.2   Automatic Test

In this section, the structure of automatic tests which have the properties described in Section 2.2 is presented. The components of an automatic test and their interaction are discussed. A way to characterize a single test is given.

The main idea of an automatic test is to apply a dedicated test signal $u(t)$ to the plant inputs which steers the plant into a specific operating region C_w which is advantageous for fault discrimination. The resulting plant output under that specific excitation is then analyzed with the goal to reject wrong fault-hypotheses. In order to achieve this goal, an automatic test consists of two main components:

1. The *input generator* which provides the test signal $u(t)$ used to excite the plant in the test-specific way.

2. The *diagnostic unit* which analyzes the input signals and output signals of the plant and provides the test result $\mathcal{H}(n+1)$ by removing wrong fault-hypotheses from the previous test result $\mathcal{H}(n)$. The diagnostic unit also provides the signal $v(t)$ which influences the input generator.

The block diagram of an automatic test for a plant which may be subject to one of the faults in the set of all possible faults $\mathcal{F}_{\mathrm{CWA}}$ is recalled in Fig. 6.2.

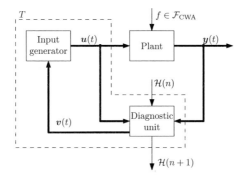

Figure 6.2: Automatic test T

The next three sections are concerned with the design of the diagnostic unit. Section 6.6 deals with the design of input generators.

6.3   Diagnostic Unit

In this section, the structure of the diagnostic unit which realizes the rejection of fault-hypotheses from the previous test result $\mathcal{H}(n)$ on the basis of the input signals $u(t)$ and the output signals $y(t)$ of the plant is explained.

The goal of the diagnostic unit is the rejection of false hypotheses on the basis of the input signals and output signals of the plant. A block diagram of such a diagnostic unit is depicted in Fig. 6.3. The main idea is to detect the presence of a specific operating region and to check the consistency of the observed I/O-pair with the nominal behavior of the plant in this operating region.

For the detection of the operating region the signal $v(t)$ which consists of validuals is used. The validuals are a new kind of signals which is introduced in Section 6.4. The consistency test is realized by a local residual $r_w(t)$. Both, the validuals and the local residual, are determined from the inputs and outputs of the plant.

The result of the detection of the operating region and the result of the consistency test are fed to a decision logic. This logic realizes the test result by removing the hypotheses $\mathcal{H}_{\mathrm{Rej}}$ from the result $\mathcal{H}(n)$ of the previous test, thus providing the test result $\mathcal{H}(n+1)$. The fault hypotheses which are removed are those which become known to be wrong if the presence of the operating region is detected and the I/O-pairs are inconsistent with the nominal behavior of the plant in this operating region.

The diagnostic unit consists of the following main blocks: the local residual generator, the operating region detection, the evaluation of the corresponding signals, and the decision logic. Each of the blocks implements a part of the above reasoning. Their purpose and interaction is detailed in the following.

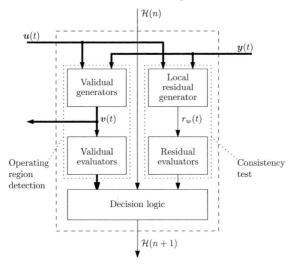

Figure 6.3: Diagnostic unit of test T

Operating region detection. This block determines whether the plant is in the operating region $\mathcal{C}_w$. For this purpose, it uses the input signal $u(t)$ and the output signal $y(t)$ of the plant to compute the signal $v(t)$. If this signal is zero, the fault-free plant is in the operating region $\mathcal{C}_w$. In this way, the signal $v(t)$ represents a measure for the distance between the present I/O-pair and the operating region $\mathcal{C}_w$ on which the test is based. The signal $v(t)$ is also fed to the input generator, thus allowing an adjustment of the input signal in order to reach the operating region $\mathcal{C}_w$. The construction of the signal $v(t)$ is detailed in Section 6.4.

Local residual generator. This block realizes the consistency check of the present I/O-pair with the nominal behavior of the plant in the operating region $\mathcal{C}_w$. It uses the input signals and output signals of the plant to determine the local residual $r_w(t)$. If the plant is in the operating region $\mathcal{C}_w$, from $r_w(t) \neq 0$ one can conclude that the present I/O-pair is inconsistent with the nominal behavior. An approach to the design of local residual generators on the basis of local structure graphs was presented in Section 5.6.

Signal evaluators. These blocks check relaxed variations of the above zero/nonzero conditions on the signals $v(t)$ and $r_w(t)$ in order to obtain Boolean variables which can be processed in a decision logic. These relaxations comprise using versions of the signals that are low-pass filtered and post-processed with characteristic curves, as well as the comparison to thresholds instead of the original zero/nonzero conditions. This is done for robustness purposes. The design of the low-pass filters, the characteristic curves and the thresholds is not discussed in this thesis as it strongly depends on the quality of the plant model used to determine the main components of the diagnostic unit.

Decision logic. In this block, the Boolean variables are used to decide whether the test is successful and the fault-hypotheses $\mathcal{H}_{\text{Rej}}$ can be removed from the previous test result $\mathcal{H}(n)$, thus providing the refined test result $\mathcal{H}(n+1)$. The design of the decision logic condensates to determining the set of fault-hypotheses $\mathcal{H}_{\text{Rej}}$. This set depends on the constraint sets used to obtain the signals $v(t)$ and $r_w(t)$. A method to determine the set $\mathcal{H}_{\text{Rej}}$ is given in Section 6.5.

The main idea of the diagnostic unit is to use the relation $v(t) = 0$ to infer on the presence of the operating region $\mathcal{C}_w$ and the relation $r_w(t) \neq 0$ to infer on the presence of a fault. If both hold true, it is possible to reject fault-hypotheses. Assuming that at each step in a test-sequence only those tests are used which can reject fault-hypotheses which are still in the hypthesis-set, a test is successful in the sense of eqn. (2.10), if

$$v(t) = \mathbf{0} \wedge r_w(t) \neq 0 \tag{6.1}$$

holds. It is unsuccessful otherwise.

In this section, the structure of the diagnostic unit of an automatic test was discussed. The purpose of its components and their interaction was presented. The subunits operating region detection and decision logic are detailed in the following sections.

6.4   Operating Region Detection

The diagnostic unit which was described in the previous section contains a subunit which is used to detect the presence of an operating region. This section presents the concept of validuals which are used in this subunit. A structure-graph-based method for the design of validuals is presented. The validual was originally introduced in [3], as a special signal which allows to infer on the presence of an operating region.

Local residuals which are obtained from the local structure graph $G|\mathcal{C}_w$ can only be used for diagnostic purposes, if the plant is in the operating region $\mathcal{C}_w$. The diagnostic unit therefore needs to detect whether this operating region is present. For this purpose, it computes a signal $\boldsymbol{v}(t)$ which indicates the presence of an operating region $\mathcal{C}_w$ if $\boldsymbol{v}(t) = \boldsymbol{0}$ holds.

In the framework of local structure graphs, an operating region $\mathcal{C}_w$ consists of constraints $c_{\mathrm{Elim},k} \in \mathcal{C}_w$ each of which correspond to an edge e_k being locally ineffective. An operating region $\mathcal{C}_w$ is present, if all the constraints $c_{\mathrm{Elim},k} \in \mathcal{C}_w$ are satisfied. For that reason, by verifying separately for each constraint $c_{\mathrm{Elim},k} \in \mathcal{C}_w$ that it is satisfied, inferring that the operating region $\mathcal{C}_w$ is present is possible. In order to verify whether a constraint $c_{\mathrm{Elim},k}$ is satisfied, only the input signals and output signals of the plant may be used. The vector $\boldsymbol{v}(t)$ can therefore consist of elements $v_k(t)$ which are computed from the plant's inputs and outputs. Each of the signals $v_k(t)$, indicates that a constraint $c_{\mathrm{Elim},k}$ is satisfied. If the number of constraints in $\mathcal{C}_w$ is p, one obtains the structure depicted in Fig. 6.4 for the operating region detection, in which the blocks VG_k provide the signals $v_k(t)$.

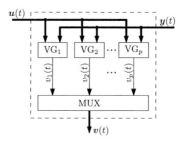

Figure 6.4: Operating region detection

With this approach, the problem of verifying the presence of a specific operating region $\mathcal{C}_w$ is reduced to finding signals $v_k(t)$, each indicating that a constraint $c_{\mathrm{Elim},k} \in \mathcal{C}_w$ is satisfied. Let the signal $\boldsymbol{v}(t)$ be a measure of the distance between the present I/O-pair and the operating region $\mathcal{C}_w$. Then $\boldsymbol{v}(t) = \boldsymbol{0}$ means that the plant is in $\mathcal{C}_w$. This leads to the idea to use $v_k(t) = 0$ as an indication that the constraint $c_{\mathrm{Elim},k}$ is satisfied. The definition of the *validual*, which was originally introduced in [3], makes use of this idea:

Definition 6.1 (Validual). *A signal $v_k(t)$ with the property that $v_k(t) = 0$ implies that the fault-free plant is in the operating region defined by the constraint $c_{\mathrm{Elim},k}$ is called* validual *for $c_{\mathrm{Elim},k}$.*

A dynamical system which computes a validual $v_k(t)$ from the input signals $\boldsymbol{u}(t)$ and the output signals $\boldsymbol{y}(t)$ of the plant is called validual generator*:*

$$v_k(t) = v_k\left(\boldsymbol{u}(t), \boldsymbol{y}(t), \frac{\mathrm{d}}{\mathrm{d}t}\right) \tag{6.2}$$

The name validual was chosen because the signal $v_k(t)$ *validates* that the fault-free plant is in the operating region $c_{\text{Elim},k}$ and the procedure to compute this signal is similar to the one used to determine a *residual*.

In the following, a method to determine validual generators is derived. The main idea is: In order to validate whether a constraint $c_{\text{Elim},k}$ is satisfied, it is sufficient to obtain the values of the variables $\text{var}^{G(\mathcal{C} \cup c_{\text{Elim},k})}(c_{\text{Elim},k})$ from the plant inputs and outputs and to inject them into the constraint $c_{\text{Elim},k}$. Under the assumption that the plant is fault-free, that is all fault variables are zero, the resulting terms of the constraint $c_{\text{Elim},k}$ form a validual. This approach is detailed in the following.

The main difficulty of the approach is to obtain the values of the unknown variables

$$\mathcal{X} \cap \text{var}^{G(\mathcal{C} \cup c_{\text{Elim},k})}(c_{\text{Elim},k}). \tag{6.3}$$

The analysis of the global structure graph of the constraint set $\mathcal{C} \cup c_{\text{Elim},k}$ allows to find a way to determine the values of these variables.

A globally structurally justconstrained set $\mathcal{C}_{\text{GSJ,Val},k}$ with a causal matching on its global structure graph that is complete w.r.t. the unknown variables $\mathcal{X} \cap \text{var}^{G(\mathcal{C} \cup c_{\text{Elim},k})}(\mathcal{C}_{\text{GSJ,Val},k})$ allows to determine the values of these unknown variables from the input and output signals of the plant. Therefore, in order to compute the values of the unknown variables which appear in the constraint $c_{\text{Elim},k}$, a justconstrained set $\mathcal{C}_{\text{GSJ,Val},k}$ with the property

$$\mathcal{X} \cap \text{var}^{G(c_{\text{Elim},k})}(c_{\text{Elim},k}) \subseteq \mathcal{X} \cap \text{var}^{G(\mathcal{C} \cup c_{\text{Elim},k})}(\mathcal{C}_{\text{GSJ,Val},k}) \tag{6.4}$$

and a causal matching that is complete w.r.t. the unknown variables can be used.

If any constraint c_{unm} is removed from a globally minimal structurally overconstrained set of constraints $\mathcal{C}_{\text{GMSO}}$, one obtains a globally structurally justconstrained constraint set:

$$\mathcal{C}_{\text{GSJ}} = \mathcal{C}_{\text{GMSO}} \setminus c_{\text{unm}}, \quad c_{\text{unm}} \in \mathcal{C}_{\text{GMSO}}. \tag{6.5}$$

This property can be used to obtain the constraint set $\mathcal{C}_{\text{GSJ,Val},k}$, which allows to determine the values of the unknown variables in $c_{\text{Elim},k}$. If one computes GMSOs

$$\mathcal{C}_{\text{GMSO},k} \subseteq (\mathcal{C} \cup c_{\text{Elim},k}) \tag{6.6}$$

and selects those which contain $c_{\text{Elim},k}$, one can obtain $\mathcal{C}_{\text{GSJ,Val},k}$ by removing the constraint $c_{\text{Elim},k}$ from the GMSO:

$$\mathcal{C}_{\text{GSJ,Val},k} = \mathcal{C}_{\text{GMSO},k} \setminus c_{\text{Elim},k}. \tag{6.7}$$

Let there be a causal matching on the global structure graph of $\mathcal{C}_{\text{GSJ,Val},k}$ which is complete w.r.t. the unknown variables occurring in the constraints in the set $\mathcal{C}_{\text{GSJ,Val},k}$. The constraints in the set $\mathcal{C}_{\text{GSJ,Val},k}$ then allow to determine the values of the unknown variables in eqn. (6.4) from the input signals and the output signals of the plant. If they are injected into the constraint $c_{\text{Elim},k}$, one obtains a validual. If there is no cycle-free matching on the global structure graph of $\mathcal{C}_{\text{GSJ,Val},k}$, it may not be possible to give a closed-form expression of the validual. If this is due to the feedback loop in a dynamical system, it may be possible to obtain a validual by forward simulation of $\mathcal{C}_{\text{GSJ,Val},k}$ which requires the initial conditions to be known.

The algorithm FINDMSO usually finds a number of GMSOs on the global structure graph of $\mathcal{C} \cup c_{\text{Elim},k}$ which satisfy the above properties. Hence, different constraint sets $\mathcal{C}_{\text{GMSO},k}$ for one single constraint $c_{\text{Elim},k}$ can be found. These are denoted with the additional subscript q:

$$\mathcal{C}_{\text{GMSO},k,q}. \tag{6.8}$$

For that reason, one can determine a number of different validuals $v_{k,q}(t)$, all of which allow to conclude that the single constraint $c_{\text{Elim},k}$ is satisfied. Of course, the algorithm FINDMSO also finds GMSOs which do not contain the constraint $c_{\text{Elim},k}$. Then eqn. (6.7) is not possible and, hence, this GMSO does not allow to determine a validual.

The above method for the determination of validuals is summed up in the following algorithm. An example of its application is given in Example 6.1:

Algorithm 4 [FindValiduals]:

Input: A plant model $\mathcal{C}$ and a constraint $c_{\text{Elim},k}$ describing an operating region.

Initialisation: An empty family of sets $\mathcal{A}_{\text{Val},k}$.

Step 1: With the global structure graph of $\mathcal{C} \cup c_{\text{Elim},k}$, compute all GMSOs

$$\mathcal{C}_{\text{GMSO},k,q} \subseteq \{\mathcal{C} \cup c_{\text{Elim},k}\}. \tag{6.9}$$

Step 2: Eliminate the GMSOs which do not contain $c_{\text{Elim},k}$.

Step 3: If there is a causal matching on the global structure graph of $\mathcal{C}_{\text{GMSO},k,q} \setminus c_{\text{Elim},k}$ that is complete w.r.t. the unknowns, add $\mathcal{C}_{\text{Val},k,q} = \mathcal{C}_{\text{GMSO},k,q}$ to $\mathcal{A}_{\text{Val},k}$.

Result: A family $\mathcal{A}_{\text{val},k}$ of constraint sets, each element $\mathcal{C}_{\text{Val},k,q} \in \mathcal{A}_{\text{Val},k}$ of which allows to determine a validual $v_{k,q}(t)$. This is done by solving $\mathcal{C}_{\text{Val},k,q} \setminus c_{\text{Elim},k}$ for the unknowns and injecting the result into the constraint $c_{\text{Elim},k}$.

Example 6.1 (Validual for $c_{\text{Elim},2}$ in Plant A). *The computation of validuals is illustrated with Plant A and the operating region $c_{\text{Elim},2}$ known from Example 5.2. The example follows the numbering of the steps in the algorithm* FINDVALIDUALS.

1. *The following GMSOs are found:*

$$\mathcal{C}_{\text{GMSO},2,1} = \{c_1, c_2, c_3, d_1\}$$
$$\mathcal{C}_{\text{GMSO},2,2} = \{c_1, c_2, c_4, c_5, d_1\}$$
$$\mathcal{C}_{\text{GMSO},2,3} = \{c_1, c_2, c_4, c_{\text{Elim},2}, d_1\}$$
$$\mathcal{C}_{\text{GMSO},2,4} = \{c_1, c_3, c_4, c_5\}$$
$$\mathcal{C}_{\text{GMSO},2,5} = \{c_1, c_5, c_{\text{Elim},2}\}$$
$$\mathcal{C}_{\text{GMSO},2,6} = \{c_2, c_3, c_4, c_5, d_1\}$$
$$\mathcal{C}_{\text{GMSO},2,7} = \{c_2, c_3, c_5, c_{\text{Elim},2}, d_1\}$$
$$\mathcal{C}_{\text{GMSO},2,8} = \{c_2, c_4, c_5, c_{\text{Elim},2}, d_1\}$$
$$\mathcal{C}_{\text{GMSO},2,9} = \{c_3, c_4, c_{\text{Elim},2}\}$$

2. *The GMSOs* $\mathcal{C}_{\mathrm{GMSO},1,1}$, $\mathcal{C}_{\mathrm{GMSO},1,2}$, $\mathcal{C}_{\mathrm{GMSO},1,4}$ *and* $\mathcal{C}_{\mathrm{GMSO},1,6}$ *are eliminated because they do not contain* $c_{\mathrm{Elim},2}$.

3. *If the constraint* $c_{\mathrm{Elim},k}$ *is removed from the remaining GMSOs, there is a causal matching which is complete w.r.t. the unknown variables which appear in the constraints in the GMSO on the global structure graph. All GMSOs are therefore added to the familiy of sets* $\mathcal{A}_{\mathrm{Val},k}$, *which results in*

$$\mathcal{A}_{\mathrm{Val},k} = \{\mathcal{C}_{\mathrm{Val},2,3}, \mathcal{C}_{\mathrm{Val},2,5}, \mathcal{C}_{\mathrm{Val},2,7}, \mathcal{C}_{\mathrm{Val},2,8}, \mathcal{C}_{\mathrm{Val},1,9}\}.$$

The validuals

$$v_{2,3}(t) : \text{(Closed form requires the solution of a differential equation.)}$$
$$v_{2,5}(t) = \frac{y_2(t)}{K_{y2}K_{\mathrm{u}}u(t)}$$
$$v_{2,7}(t) = \frac{TK_{y1}y_2(t)}{K_{y2}(y_1(t) - T\frac{\mathrm{d}}{\mathrm{d}t}y_1(t))}$$
$$v_{2,8}(t) : \text{(Closed form requires the solution of a differential equation.)}$$
$$v_{2,9}(t) = \frac{K_{\mathrm{M}}}{K_{y1}}y_1(t)$$

can be obtained by computing the unknown variables with $\mathcal{C}_{\mathrm{Val},k,q} \setminus c_{\mathrm{Elim},2}$ *and injecting the result into the constraint* $c_{\mathrm{Elim},2}$. *The validuals indicate the presence of the operating region* $c_{\mathrm{Elim},2}$, *if they are zero and the plant is not faulty.*

In order to compute the validual the plant is assumed to be fault-free. For faulty plants, a vanishing validual does not necessarily imply that the plant is in the corresponding operating region. To the contrary, faults which occur in the constraints used to determine the validual generator may lead to a zero validual, although the plant is not in the operating region $c_{\mathrm{Elim},k}$. Faults which do not occur in the constraint set which is used to determine the validual do not have this effect. This reasoning is used in Section 6.5 to determine the faults which a particular test is sensitive to.

The algorithm FINDVALIDUALS usually does not find only one, but a number of validuals $v_{k,q}(t)$ for one edge e_k which is locally ineffective in $c_{\mathrm{Elim},k}$. An operating region can therefore be detected by a large number of validual-combinations. In Section 6.5 it is seen that the choice of a validual $v_{k,q}(t)$ and, therefore, the choice of an operating region detection influences the sensitivity of a test significantly. This is used in Section 6.7 in order to determine tests with desired fault-discrimination properties.

In this section, a method to infer on the presence of an operating region was presented. This method uses the plant's input signals and output signals to compute a special type of signal called validual. Each validual corresponds to a constraint in the description of an operating region and allows to verify whether this constraint is satisfied. An algorithm was given to obtain validuals for a given constraint of an operating region.

6.5   Decision Logic Design

In this section, an approach to the design of the decision logic used in the diagnostic unit is developed. The approach is based on an analysis of the question which faults may lead to a nonzero local residual if the validuals which indicate the presence of the corresponding local structure are zero. The result of this analysis is used together with the Single-Fault-Assumption and the Closed-World-Assumption to design a decision logic which rejects wrong fault-hypotheses, thus realizing the outcome of an automatic test.

Regarding the validuals $\boldsymbol{v}(t)$ and the local residual $r_w(t)$, one can distinguish the following situations: Either the signal is zero or the signal is nonzero. Hence, during the course of the test, the four cases given in Tab. 6.1 may occur.

Table 6.1: Signal-cases in the diagnostic unit

		$r_w(t)$	
		$= 0$	$\neq 0$
$\boldsymbol{v}(t)$	$\neq \boldsymbol{0}$	Case A	Case B
	$= \boldsymbol{0}$	Case C	Case D

The two cases A and B do not allow to conclude on the fault-state of the plant. This is the case because not all the validuals which indicate the presence of the operating region in which $r_w(t)$ is a local residual are zero. Hence, the signal $r_w(t)$ is not the result of a consistency test. In Case C, the validuals indicate the presence of the operating region in which $r_w(t)$ is a local residual. However, since $r_w(t) = 0$, the input and output signals observed at the plant are consistent with the behavior of the plant in this operating region. Therefore, no conclusion on the presence of a fault is possible. Case D can be used for the diagnosis of the plant. All the validuals are zero. This indicates the presence of the operating region in which $r_w(t)$ is a local residual. Since $r_w(t)$ is nonzero, the inconsistency of the observed I/O-pair with the plant behavior in the operating region is shown. Therefore a fault is known to be present. The information which faults may lead to this inconsistency can be used to reject wrong fault-hypotheses. The following is therefore concerned with a method which allows to determine these faults.

6.5.1   Test Sensitivity

A test is successful if wrong fault-hypotheses are rejected. Using the exclusion principle, this rejection is only possible if the presence of other faults is detected. This requires all the validuals indicating the presence of an operating region which correspond to a specific local structure to be zero and a local residual in this operating region to be nonzero (Case D). If this is the case, one can infer on the presence of a fault. Typically, only a subset

$$\mathcal{F}_{\text{Sens}} \subseteq \mathcal{F}_{\text{CWA}} \qquad (6.10)$$

of all possible faults may be the reason for the nonzero local residual. This allows to refine the diagnostic result. In order to determine the fault-hypotheses $\mathcal{H}_{\mathrm{Rej}}$ that can be rejected by the test, the subset $\mathcal{F}_{\mathrm{Sens}}$ needs to be known.

In Definition 6.1, the plant was assumed to be fault-free in order to infer from a zero validual on the presence of an operating region. This assumption may not hold. As a matter of fact if all the validuals $v_k(t)$ indicating a specific operating region $\mathcal{C}_w$ are zero, and a local residual $r_{w,s}(t)$ in this operating region is nonzero, either of the two following situations may have occurred:

- **The plant *is* in the operating region $\mathcal{C}_w$.** If the plant actually is in the operating region $\mathcal{C}_w$, and the local residual $r_{w,s}(t)$ is nonzero, the observed I/O-pair is inconsistent with the set of constraints $\mathcal{C}_{\mathrm{Res},w,s}$ from which the local residual was obtained. Only the fault variables in that constraint set $\mathcal{C}_{\mathrm{Res},w,s}$ can be the reason for the inconsistency. These fault variables are

$$\mathcal{F}_{\mathrm{Con}} = \mathcal{F} \cap \mathrm{var}^{G(\mathcal{C})|\mathcal{C}_w}\left(\mathcal{C}_{\mathrm{Res},w,s}\right). \tag{6.11}$$

- **The plant *is not* in the operating region $\mathcal{C}_w$.** In the fault-free case, the plant can be guaranteed to be in the operating region $\mathcal{C}_w$ if all corresponding validuals $v_k(t)$ are zero. However, a fault in the plant may lead to the validuals being zero, although the plant is actually *not* in the operating region $\mathcal{C}_w$. This may happen if the part of the plant which is used to infer on the presence of an operating region is faulty, c.f. Fig. 5.3. Only those fault variables may have an impact on the validuals that occur in the constraints which are used to determine the validuals. For that reason, a validual $v_{k,q}(t)$ may only be zero although the corresponding constraint $c_{\mathrm{Elim},k}$ is not satisfied if a fault in the constraint set $\mathcal{C}_{\mathrm{Val},k,q}$ has occurred. If there is only one validual $v_{k,q}(t)$ in the test, these are the faults in the set

$$\mathcal{F}_{\mathrm{Ste},k} = \mathcal{F} \cap \mathrm{var}^{G(\mathcal{C}\cup c_{\mathrm{Elim},k})}\left(\mathcal{C}_{\mathrm{Val},k,q}\right). \tag{6.12}$$

If there is more than one constraint in $\mathcal{C}_w$, more than one validual is necessary in order to infer on the presence of the operating region $\mathcal{C}_w$. All constraint sets used to determine a validual used in the test may be subject to faults. All these faults may lead to a zero validual although the plant is not in the operating region $\mathcal{C}_w$. For that reason, also the faults

$$\mathcal{F}_{\mathrm{Ste}} = \bigcup_k \mathcal{F}_{\mathrm{Ste},k} = \bigcup_k \mathcal{F} \cap \mathrm{var}^{G(\mathcal{C}\cup c_{\mathrm{Elim},k})}\left(\mathcal{C}_{\mathrm{Val},k,q}\right) \tag{6.13}$$

may be present if all validuals indicating the presence of the operating region $\mathcal{C}_w$ are zero and a local residual in $\mathcal{C}_w$ is nonzero.

In summary, with the Single-Fault-Assumption one can conclude that either one of the faults in $\mathcal{F}_{\mathrm{Ste}}$ or one of the faults in $\mathcal{F}_{\mathrm{Con}}$ may be present. This leads to the following theorem:

Theorem 6.1 (Sensitivity). *If an automatic test with a diagnostic unit according to Section 6.3 satisfies the detection properties, that is all validuals $v_{k,q}(t)$ indicating the presence of the operating region $\mathcal{C}_w$ are zero and the local residual $r_{w,s}(t)$ in this operating region is nonzero, one of the faults in the set $\mathcal{F}_{\mathrm{Sens}}$ is present:*

$$v_{k,q}(t) = 0 \ \forall \ k \quad and \quad r_{w,s}(t) \neq 0 \quad \Rightarrow \quad \exists \ f_i \neq 0, \ f_i \in \mathcal{F}_{\mathrm{Sens}} = \mathcal{F}_{\mathrm{Con}} \cup \mathcal{F}_{\mathrm{Ste}} \quad (6.14)$$

with

$$\mathcal{F}_{\mathrm{Con}} = \mathcal{F} \cap \mathrm{var}^{G(\mathcal{C})|\mathcal{C}_w} \left(\mathcal{C}_{\mathrm{Res},w,s} \right) \quad (6.15)$$

$$\mathcal{F}_{\mathrm{Ste}} = \bigcup_k \mathcal{F} \cap \mathrm{var}^{G(\mathcal{C} \cup \mathcal{C}_{\mathrm{Elim}},k)} \left(\mathcal{C}_{\mathrm{Val},k,q} \right). \quad (6.16)$$

Note that in this context, the term *sensitivity* refers to a binary statement which is not obtained from a differentiation: A test is either sensitive to a fault or not.

The result in Theorem 6.1 is analogous to a result in [81], where a plant is steered into predefined discrete modes in which it is diagnosed under the assumption that steering the plant may not fail. This corresponds to $\mathcal{F}_{\mathrm{Ste}} = \emptyset$. The effect that steering the plant in the desired way unsuccessfully may cause a wrong diagnostic result is described in [81], but a detailed explanation is not given. Example 6.2 shows the application of Theorem 6.1.

Example 6.2 (Test for Plant A). *In Example 6.1, the constraint set $\mathcal{C}_{\mathrm{Val},2,7}$ was used to determine the validual $v_{2,7}(t)$ for the operating region $\mathcal{C}_2$. In Example 5.6, the constraint set $\mathcal{C}_{\mathrm{Res},2,1} = \mathcal{C}_{\mathrm{LMSO},2,1}$ was used to determine the local residual $r_{2,1}(t)$. With Theorem 6.1, one obtains*

$$\mathcal{F}_{\mathrm{Ste}} = \{f_2, f_3, f_5\}$$

and

$$\mathcal{F}_{\mathrm{Con}} = \{f_1, f_2, f_3\}$$

which results in

$$\mathcal{F}_{\mathrm{Sens}} = \{f_1, f_2, f_3, f_5\}. \quad (6.17)$$

This result means that if $v_{2,7}(t) = 0$ and $r_{2,1}(t) \neq 0$ hold true, one of the faults in $\mathcal{F}_{\mathrm{Sens}}$ is present.

The equations (6.14)-(6.16) explain, why the choice of an operating region is a crucial step in the design of a test. Choosing small operating regions usually entails a small number of faults which may lead to the local residual being nonzero if the plant actually is in the operating region. In this case, the set $\mathcal{F}_{\mathrm{Con}}$ is small. However, a large number of validuals may be necessary to validate the presence of the operating region. This in turn leads to a large set $\mathcal{F}_{\mathrm{Ste}}$. The faults $\mathcal{F}_{\mathrm{Sens}}$, to which a test is sensitive, consist of both, $\mathcal{F}_{\mathrm{Con}}$ and $\mathcal{F}_{\mathrm{Ste}}$. Hence, the operating region must be chosen carefully in order to obtain a test with the desired sensitivity.

In the suggested way, the difficult problem of steering a potentially faulty system was reduced to the problem of detecting the operating region in a potentially faulty plant. The further reduction to detecting the presence of an operating region in a fault-free plant and considering the faults which may prohibit this allows to determine the set of faults to which a diagnostic unit is sensitive. In this way, the necessity of an operating region detection which detects the actual presence of an operating region in a faulty plant can be avoided.

6.5.2   Hypothesis Rejection

An automatic test rejects wrong fault-hypotheses. In the following, additionally to the Single-Fault-Assumption, the Closed-World-Assumption is used in order to obtain the set of fault-hypotheses which can be rejected, if the validuals are zero and a corresponding local residual is nonzero. The Closed-World-Assumption is the assumption that all faults which are relevant for the diagnosis of the plant are known and modeled as the fault variables $\mathcal{F}_{\mathrm{CWA}}$ occurring in the constraints in the set $\mathcal{C}$ of the analytical model. In the context of service diagnosis, the latter assumption is justified for two reasons: First, typical faults are usually known, c.f. Section 3.3.3. Second, for each constraint in the plant model, a general fault may be modeled by simply adding a fault variable to the constraint. Then each general fault corresponds to a constraint which is the same as considering a fault being the violation of a constraint. It is then possible to associate every constraint to a number of components which allows to determine the faulty component on the basis of the faulty constraints. However, this approach is conservative because then local structure graphs, in which a fault variable is not connected to the rest of the local structure graph, do not exist.

By exclusion principle, the Single-Fault-Assumption, the Closed-World-Assumption and the result in Theorem 6.1 allow to reject all fault-hypotheses corresponding to the fault variables in the set

$$\mathcal{F}_{\mathrm{Rej}} = \mathcal{F}_{\mathrm{CWA}} \setminus \mathcal{F}_{\mathrm{Sens}}. \tag{6.18}$$

These fault-hypotheses are the hypotheses that the fault variables in $\mathcal{F}_{\mathrm{Rej}}$ are nonzero. Therefore, if all the validuals are zero and the corresponding local residual is nonzero, a diagnostic unit can reject the fault-hypotheses in the set

$$\mathcal{H}_{\mathrm{Rej}} = \bigcup h_i : f_i \neq 0 \ \forall \ f_i \in \{\mathcal{F}_{\mathrm{CWA}} \setminus \mathcal{F}_{\mathrm{Sens}}\}. \tag{6.19}$$

The outcome of the test is therefore

$$\mathcal{H}(n+1) = \begin{cases} \mathcal{H}(n) \setminus \mathcal{H}_{\mathrm{Rej}} & \text{if } \boldsymbol{v}(t) = \mathbf{0} \wedge r_{w,s}(t) \neq 0 \\ \mathcal{H}(n) & \text{otherwise.} \end{cases} \tag{6.20}$$

The above result allows to construct a decision logic which realizes the output $\mathcal{H}(n+1)$ of the diagnostic unit of an automatic test which can be used in the framework of service diagnosis, c.f. Section 2.2. Example 6.3 shows the application of the above result.

Example 6.3 (Hypothesis rejection for Plant A).   *With the set of all possible faults* $\mathcal{F}_{\mathrm{CWA}} = \{f_1, f_2, f_3, f_4, f_5\}$, *and the result of the previous example,* $\mathcal{F}_{\mathrm{Sens}} = \{f_1, f_2, f_3, f_5\}$, *the hypothesis-set which can be rejected if the validual $v_{2,7}(t)$ is zero and the local residual $r_{2,1}(t)$ is nonzero is*

$$\mathcal{H}_{\mathrm{Rej}} = \{h_4\}. \tag{6.21}$$

Assuming that the result of the previous test was $\mathcal{H}(n) = \{h_4, h_5\}$, the test result is

$$\mathcal{H}(n+1) = \{h_5\}, \tag{6.22}$$

if

$$v_{2,7}(t) = 0 \ \wedge \ r_{2,1}(t) \neq 0 \tag{6.23}$$

holds true. Note that a result of the analysis of the global structural diagnosability properties of Plant A was that the faults f_4 and f_5 form a group of faults that are not globally structurally discriminable, c.f. Example 4.8. The example at hand therefore shows, how tests based on local residuals and validuals allow to discriminate faults that are not distinguishable using global residuals found by the analysis of the global structure graph.

This section developed the diagnostic unit of an automatic test. The unit consists of a local residual generator, a detection of the operating region on the basis of validual generators and a decision logic which allows the rejection of wrong fault-hypotheses. In that way, the presented approach allows the integration of tests with the resulting diagnostic unit into the framework of service diagnosis, c.f. Chapter 2.

6.6   Input Generator Design

This section develops a major contribution of this thesis, namely a method for determining the input generator of an automatic test. It starts from the idea that for a successful test all the validuals indicating the presence of a specific operating region need to be zero. The constraint sets used to determine the validuals allow to find constraints on input signals ensuring that the validuals are zero. Three special cases of the realization of the input generator are distinguished: pure feedforward controller, pure feedback controller and a control law computing the input signals from the output signals of the plant.

According to the previous section, a test can only be successful if all the validuals $v_k(t)$ indicating the presence of the operating region $\mathcal{C}_w$ are zero. Hence, the plant's input signals $\boldsymbol{u}(t)$ and output signals $\boldsymbol{y}(t)$ need to take on values which ensure this. The task of the input generator is therefore to provide test signals $\boldsymbol{u}(t)$ which, if applied to the plant's inputs, lead to output signals $\boldsymbol{y}(t)$, such that the validuals are zero:

$$\boldsymbol{u}(t): \ v_k\left(\boldsymbol{u}(t), \boldsymbol{y}(t), \frac{\mathrm{d}}{\mathrm{d}t}\right) = 0 \ \forall \ k. \tag{6.24}$$

Equation (6.24) and the fact that the validual generators $v_k(\boldsymbol{u}(t), \boldsymbol{y}(t), \frac{\mathrm{d}}{\mathrm{d}t})$ are known, motivate the idea to set the validual generators zero and to solve for the input variables $u \in \mathcal{U}$ in order to obtain the test signals. The validual may contain derivatives w.r.t. time of the input and output signals of the plant. Therefore, the solution for the input variables may contain these derivatives as well. In that case the test signal is described by a differential equation.

Because the validual generators $v_{k,q}(\boldsymbol{u}(t), \boldsymbol{y}(t), \frac{\mathrm{d}}{\mathrm{d}t})$ are obtained from the constraint sets $\mathcal{C}_{\mathrm{Val},k,q}$, these constraint sets also characterize the input generator which provides the test signal. In particular, because the sets $\mathcal{C}_{\mathrm{Val},k,q}$ form relations of input variables $u \in \mathcal{U}$ and output variables $y \in \mathcal{Y}$, three cases can be distinguished:

Only input signals in the validual generators. If there are only input variables in the known variables of the constraint sets $\mathcal{C}_{\mathrm{Val},k,q}$, the validual generators contain only input signals of the plant and their derivatives w.r.t. time:

$$\mathcal{Y} \cap \mathrm{var}^{G(\mathcal{C} \cup \mathcal{C}_{\mathrm{Elim}},k)}\left(\mathcal{C}_{\mathrm{Val},k,q}\right) = \emptyset \; \forall \; k$$
$$\Rightarrow v_{k,q}(t) = v_{k,q}\left(\boldsymbol{u}(t), \frac{\mathrm{d}}{\mathrm{d}t}\right) \; \forall \; k. \tag{6.25}$$

If the constraints $0 = v_{k,q}(\boldsymbol{u}(t), \frac{\mathrm{d}}{\mathrm{d}t})$, are solved for the input signal $\boldsymbol{u}(t)$, the result cannot depend upon the plant's output. In that case, the input generator is a feedforward controller. Considering only the operating region detection of the diagnostic unit, a signal flow according to the block diagram in Fig. 6.5 is obtained.

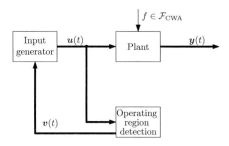

Figure 6.5: Input generator realized by a feedforward controller

Input signals and output signals in a validual generator. If both input and output variables appear in one of the constraint sets $\mathcal{C}_{\mathrm{Val},k,q}$, the corresponding validual generator contains both input signals and output signals of the plant:

$$\exists \; k : \; \mathcal{Y} \cap \mathrm{var}^{G(\mathcal{C} \cup \mathcal{C}_{\mathrm{Elim}},k)}\left(\mathcal{C}_{\mathrm{Val},k,q}\right) \neq \emptyset \; \wedge \; \mathcal{U} \cap \mathrm{var}^{G(\mathcal{C} \cup \mathcal{C}_{\mathrm{Elim}},k)}\left(\mathcal{C}_{\mathrm{Val},k,q}\right) \neq \emptyset$$
$$\Rightarrow \exists \; k : \; v_{k,q}(t) = v_{k,q}\left(\boldsymbol{u}(t), \boldsymbol{y}(t), \frac{\mathrm{d}}{\mathrm{d}t}\right). \tag{6.26}$$

If the constraints $0 = v_{k,q}(\boldsymbol{u}(t), \boldsymbol{y}(t), \frac{\mathrm{d}}{\mathrm{d}t})$ are solved for the input variables, one obtains expressions that contain signal outputs, their derivatives w.r.t. time and derivatives w.r.t. time of the input signals. In that case the input generator is realized by a control law.

A similar result can be found in [105]: Faults in the system matrix and the input matrix of a linear dynamical system can be distinguished by applying test signals to the system inputs which are obtained by a feedback controller. This analogy is described in Remark 6.1 in detail.

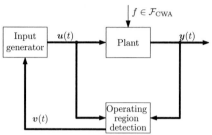

Figure 6.6: Input generator realized by a control law

Only output signals in the validual generators. If there are only output variables in the known variables of the constraint sets $\mathcal{C}_{\text{Val},k,q}$, the validual generators contain only output signals of the plant:

$$\mathcal{U} \cap \text{var}^{G(\mathcal{C} \cup c_{\text{Elim},k})}\left(\mathcal{C}_{\text{Val},k,q}\right) = \emptyset \ \forall \ k$$
$$\Rightarrow v_{k,q}(t) = v_{k,q}\left(\boldsymbol{y}(t), \frac{\mathrm{d}}{\mathrm{d}t}\right) \ \forall \ k. \tag{6.27}$$

In that case, it is not possible to set the validual generator to zero and to solve the expression for the input variables. Nevertheless, the claim that the test signals applied to the plant need to make the validuals zero, holds. Therefore the plant outputs need to be steered in a way that makes the validuals zero. This can be reached by robust feedback controllers, the design of which is left to a control engineer.

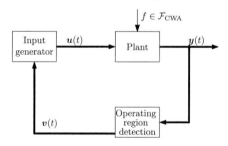

Figure 6.7: Input generator realized by a feedback controller

The above approach to the design of input generators does not always result in an unambiguous solution for the input signals. As a matter of fact, the approach finds *constraints* on the input signals. Three cases can be distinguished:

- **More than one solution for $u(t)$:** Particularly if the operating region $\mathcal{C}_w$ contains only few constraints and, therefore, the operating region detection contains only few validuals, the approach to determine the input generator may not result in an unambiguous solution for the input signals. In that case the remaining degree of freedom can be used in order to maximize the impact of a given fault on the local residual.

- **Unambiguous solution for $u(t)$:** The approach may result in an unambiguous solution for $u(t)$. In that case only one particular input signal $u(t)$ leads to all the validuals being zero.

- **No solution for $u(t)$:** If the operating region $\mathcal{C}_w$ consists of a large number of constraints, and the plant has only few inputs, it may happen that there is no solution leading to all the validuals being zero. This can be the case if the operating region is ill-defined. This means that $\mathcal{C}_w \cup \mathcal{C}$ does not have a solution for the unknown variables which appear in the constraints in the set $\mathcal{C}_w$ for arbitrary input signals $u(t)$ and output signals $y(t)$. This is for instance the case if it is not possible to steer the plant from the given initial conditions into the operating region $\mathcal{C}_w$.

The following gives an example of an input generator for Plant A.

Example 6.4 (Input Generator for Plant A). *If the diagnostic unit of a test for Plant A consists of a local residual and the validual $v_{2,7}(t)$ obtained from $\mathcal{C}_{\mathrm{Val},2,7}$, c.f. Example 6.3, one finds*

$$\mathcal{U} \cap \mathrm{var}^{G(\mathcal{C} \cup \mathcal{C}_{\mathrm{Elim}},2)} \left(\mathcal{C}_{\mathrm{Val},2,7} \right) = \mathcal{U} \cap \{y_1, y_2\} = \emptyset \tag{6.28}$$

and the input generator can be realized by a feedback controller. This feedback controller needs to be designed in a way that it leads to $y_2(t) = 0$ and $y_1(t) \neq T\frac{\mathrm{d}}{\mathrm{d}t}y_1(t)$ because then $v_{2,7}(t) = 0$ holds.

In this section, a method to determine input generators for automatic tests was described. The approach uses the idea that for a successful test the validuals which indicate the operating region in which the signal $r_{w,s}(t)$ is a local residual need to be zero. Setting the validual generators to zero and solving the equation for the input variables yields constraints on the input signal. Any input signal which satisfies these constraints is a valid test signal. The remaining degrees of freedom in the design of the test signals can be used in order to maximize the impact of faults.

6.7 Defining All Tests

In this section, a method for the design of tests with desired properties, originally published in [4], is given. It computes all theoretically feasible tests which consist of combinations of validual generators and local residual generators. For this purpose, all operating regions, which correspond to a specific local structure, are considered. The result in Theorem 6.1 is used to select tests with desired hypotheses-rejection properties.

The design of an automatic test can be divided into two separate steps:

1. The design of a diagnostic unit which provides the signal $v(t)$ and the refined diagnostic result $\mathcal{H}(n+1)$.

2. The design of an input generator so that the validuals which form $v(t)$ become zero.

In the following, an approach to the design of the diagnostic unit is described. Once the diagnostic unit of an automatic test is known, the methods described in Section 6.6 can be used to determine an appropriate input generator for this test. In the design of a diagnostic unit according to Section 6.3, there are three degrees of freedom which influence the hypotheses-rejection properties of the overall test. These are: the choice of an operating region, the choice of validuals which are used to infer on the presence of the operating region and the choice of a local residual which is used for the consistency check.

An intuitive approach to determine a diagnostic unit consists of computing all combinations of the possible choices in a first step, and to exclude tests that cannot be successful in a second step. Then, tests with desired fault-hypotheses-rejection properties are selected in a last step. The three different degrees of freedom in the design of an automatic test are explained more detailed in the following.

The choice of an operating region $\mathcal{C}_w$. Different operating regions $\mathcal{C}_w$ lead to different local structure graphs $G|\mathcal{C}_w$ which in turn result in different local residuals. Also, they require different validuals to indicate their presence. With the result in Theorem 6.1, it is possible to conclude that the choice of an operating region influences the hypothesis rejection properties of an automatic test. It is therefore a degree of freedom in the design of such tests.

If all locally ineffective edges are stored in the set $\mathcal{E}_{\mathrm{Elim}}$ and the corresponding constraints are stored in the set $\mathcal{C}_{\mathrm{Elim}}$, all constraint sets in the power set of $\mathcal{C}_{\mathrm{Elim}}$ are candidates for operating regions. Such a candidate is denoted by $\mathcal{C}_{\mathrm{CFO}}$. For two reasons it needs to be verified whether the constraint sets $\mathcal{C}_{\mathrm{CFO}}$ are valid operating regions: First, the constraints in a constraint set $\mathcal{C}_{\mathrm{CFO}} \in 2^{\mathcal{C}_{\mathrm{Elim}}}$ may contradict each other. Second, a constraint set $\mathcal{C}_{\mathrm{CFO}}$ may contradict the constraint set $\mathcal{C}$ which describes the plant behavior. In both cases it is not possible to steer the plant in a way that the constraints in $\mathcal{C}_{\mathrm{CFO}}$ are satisfied simultaneously. Then, the constraint set $\mathcal{C}_{\mathrm{CFO}}$ does not represent a valid operating region. By investigating all constraint sets

$$\mathcal{C} \cup \mathcal{C}_{\mathrm{CFO}}, \ \mathcal{C}_{\mathrm{CFO}} \in 2^{\mathcal{C}_{\mathrm{Elim}}} \tag{6.29}$$

for contradiction, one can find the family of sets $\mathcal{A}_{\mathrm{VOR}} = \{\mathcal{C}_1, \mathcal{C}_2, ...\}$ which contains all valid operating regions. If the valid operating regions are known, all local structure graphs

$$G|\mathcal{C}_w, \ \mathcal{C}_w \in \mathcal{A}_{\mathrm{VOR}} \tag{6.30}$$

can be considered in order to obtain a maximum number of possible automatic tests.

The choice of validual generators $v_{k,q}(u(t), y(t), \frac{\mathrm{d}}{\mathrm{d}t})$. In order to infer on the presence of an operating region, a validual $v_{k,q}(t)$ for each constraint $c_{\mathrm{Elim},k} \in \mathcal{C}_w$ is used. With the algorithm FINDVALIDUALS for each constraint $c_{\mathrm{Elim},k}$, a family $\mathcal{A}_{\mathrm{Val},k}$ of constraint sets is found. Each of its elements $\mathcal{C}_{\mathrm{Val},k,q}$ allows to determine a validual generator $v_{k,q}(\boldsymbol{u}(t), \boldsymbol{y}(t), \frac{\mathrm{d}}{\mathrm{d}t})$. All the validuals $v_{k,q}(t)$ they provide allow to determine whether the constraint $c_{\mathrm{Elim},k}$ is satisfied. An arbitrary validual generator can therefore be chosen from the validual generators provided by FINDVALIDUALS to indicate that $c_{\mathrm{Elim},k}$ is satisfied. In general, an operating region consists of more than one constraint. Then the vector valued signal $\boldsymbol{v}(t)$ which consists of the validuals $v_{k,q}(t)$ is used to infer on the presence of an operating region $\mathcal{C}_w$. Because for each constraint $c_{\mathrm{Elim},k}$ the above reasoning holds, all combinations of validuals may be used to construct the signal $\boldsymbol{v}(t)$. Each of these combinations is based on different GMSOs $\mathcal{C}_{\mathrm{Val},k,q}$. With Theorem 6.1 one can conclude that different combinations of these GMSOs which can all be used to determine the signal $\boldsymbol{v}(t)$ have a different impact on the hypotheses-rejection properties of the resulting test. The choice of a combination of validual generators therefore represents an important degree of freedom which needs to be considered in the design of an automatic test.

The choice of a local residual generator $r_{w,s}(\boldsymbol{u}(t), \boldsymbol{y}(t), \frac{\mathrm{d}}{\mathrm{d}t})$. If the plant is in the operating region $\mathcal{C}_w$, the local structure $G|\mathcal{C}_w$ holds. The algorithm FINDLOCALRESID-UALS applied to the local structure graph $G|\mathcal{C}_w$ results in a number of constraint sets $\mathcal{C}_{\mathrm{Res},w,s}$ and the corresponding local residual generators $r_{w,s}(\boldsymbol{u}(t), \boldsymbol{y}(t), \frac{\mathrm{d}}{\mathrm{d}t})$. According to Theorem 6.1, the hypotheses which can be rejected if an automatic test is successful depend upon the local residual generator. Consequently, the choice of an LMSO $\mathcal{C}_{\mathrm{Res},w,s}$ which allows to determine a local residual generator $r_{w,s}(\boldsymbol{u}(t), \boldsymbol{y}(t), \frac{\mathrm{d}}{\mathrm{d}t})$ is a further degree of freedom in the design of automatic test.

Combining a feasible operating region with arbitrary validuals for the constraints $c_{\mathrm{Elim},k}$ which represent the operating region $\mathcal{C}_w$ and an arbitrary local residual in this operating region one obtains the diagnostic unit of an automatic test. This forms the basis of the method for the design of tests.

It is possible that for a diagnostic unit which is obtained by an arbitrary combination of validuals and a local residual, the case

$$\boldsymbol{v}(t) = \boldsymbol{0} \ \wedge \ r_{w,s}(t) \neq 0 \tag{6.31}$$

never occurs. This is in particular the case if the same sets of constraints are used in order to obtain the local residual generator and a validual generator:

$$\exists \ k, c_{\mathrm{Elim},k} \in \mathcal{C}_w : \ \mathcal{C}_{\mathrm{Val},k,q} = \mathcal{C}_{\mathrm{Res},w,s}. \tag{6.32}$$

If eqn. (6.32) holds for some q and s, the following situation occurs: if on one hand $v_{k,q}(t) = 0$ holds, the observed I/O-pair is consistent with $\mathcal{C}_{\mathrm{Val},k,q}$. Because of eqn. (6.32), the I/O-pair is also consistent with $\mathcal{C}_{\mathrm{Res},w,s}$ and, therefore, the local residual is zero, c.f. Case C in Tab. 6.1. Then, the test is unsuccessful.

If on the other hand $r_{w,s}(t) \neq 0$, the I/O-pair is inconsistent with $\mathcal{C}_{\mathrm{Res},w,s}$. Then, because eqn. (6.32) holds, the I/O-pair is also inconsistent with $\mathcal{C}_{\mathrm{Val},k,q}$ and the validual $v_{k,q}(t)$ is nonzero, c.f. Case A. This also means that the test was unsuccessful.

This reasoning shows that although, from the point of view of the logical conclusion, tests with the property eqn. (6.32) are valid, they cannot be successful. These tests are therefore not useful for service diagnosis. For this reason, they need to be excluded from the set of combinations obtained in the design step of the diagnostic unit.

The following algorithm sums up the procedure to determine automatic tests.

Algorithm 5 [FindTests]:

Input: The analytical plant model with the constraint set $\mathcal{C}$.

Step 1: Determine the set of all locally ineffective edges $\mathcal{E}_{\mathrm{Elim}} \subseteq \mathcal{E}$ with Definition 5.2 and the corresponding constraints $\mathcal{C}_{\mathrm{Elim}}$ under which the edges are locally ineffective.

Step 2: Determine all valid operating regions represented by the family of sets

$$\mathcal{A}_{\mathrm{VOR}} \subseteq 2^{\mathcal{C}_{\mathrm{Elim}}}. \tag{6.33}$$

Step 3: For each operating region $\mathcal{C}_w \in \mathcal{A}_{\mathrm{VOR}}$, do:

Step 4: Determine $\mathcal{A}_{\mathrm{Res},w}$ with FINDLOCALRESIDUALS.

Step 5: For each constraint in $c_{\mathrm{Elim},k} \in \mathcal{C}_w$, determine $\mathcal{A}_{\mathrm{Val},k}$ with FINDVALIDUALS.

Step 6: For each combination of the constraint sets $\mathcal{C}_{\mathrm{Val},k,q} \in \mathcal{A}_{\mathrm{Val},k}$, do:

Step 7: Store the combination of the constraint sets $\mathcal{C}_{\mathrm{Val},k,q}$ and the constraint set $\mathcal{C}_{\mathrm{Res},w,s} \in \mathcal{A}_{\mathrm{Res},w}$, if the property eqn. (6.32) does not hold.

Step 8: Use the result of Theorem 6.1 to determine which hypotheses a test may reject if its diagnostic unit is obtained from the combination of the constraint sets $\mathcal{C}_{\mathrm{Val},k,q}$ and the constraint set $\mathcal{C}_{\mathrm{Res},w,s}$.

Result: A list of constraint sets which allow to determine a diagnostic unit and the hypotheses $\mathcal{H}_{\mathrm{Rej}}$ it may reject.

This method has two major advantages: First, the structural analysis allows to select an automatic test according to its hypotheses-rejection properties *before* actually computing the validuals and the local residual. For this purpose, it makes use of an abstraction of a detailed model in order to cope with large-scale systems.

Second, a test does usually not require a complete parameterized model of the plant like the observer-based approaches in [11] and [12]. Only a subset of the constraints governing the system-behavior needs to be known. This reduces the design effort significantly, particularly in terms of identification of the parameters.

Generally, the more edges are removed from the global structure by choice of an operating region, the fewer faults influence a local residual. However, the larger the cardinality of the set $\mathcal{E}_w$ is, the more validuals occur in the test. According to Theorem 6.1 a test is sensitive to both, faults influencing the local residual *and* the validuals. Therefore, tests obtained by local structures with a large cardinality $|\mathcal{E}_w|$ are not necessarily the ones sensitive to the fewest faults.

In this section, an approach to the design of automatic tests was suggested. The degree of freedom in the design of an automatic test is threefold: namely the choice of an operating region, the choice of a combination of validual generators and the choice of a local residual generator. Based on this observation, it was suggested to compute the constraint sets defining all theoretically feasible tests. The approach to the selection of automatic tests consists of excluding those tests that cannot be successful. Then, tests with desired discrimination properties can be chosen. All of this is done using global and local structural models of the plant.

6.8   Special Case: Fault Hiding

In this section, a special kind of automatic tests which may result from the methods described in Section 6.2 - Section 6.7 is presented. The operating region used for these tests corresponds to a local structure graph, in which the edges between a fault variable and the rest of the structure graph are locally ineffective. In the corresponding operating region, the fault has no impact on the entire plant behavior – the fault is hidden. A dedicated method of finding such tests can be found in [2] and [5] where it is illustrated with the example of an SI engine and a throttle valve, respectively.

In a specific operating region, a particular fault f_i may not have any impact on the behavior of the plant. If a plant is brought into such an operating region, all the I/O-pairs are the same, regardless of the value of the fault variable. That is, when the plant is excited with the specific input signal $\boldsymbol{u}(t)$, the plant will answer with the same output signal $\boldsymbol{y}(t)$ independently of the presence of the fault f_i. In that case, the fault is said to be hidden. These specific signals form I/O-pairs which belong to the nominal behavior $\mathcal{B}(\mathcal{C}_0)$ and the faulty behavior $\mathcal{B}(\mathcal{C}_{f_i})$ of the plant:

$$\mathcal{B}(\mathcal{C}_0) \cap \mathcal{B}(\mathcal{C}_{f_i}) \neq \emptyset. \tag{6.34}$$

Obviously, if the plant is known to be in the specific operating region, and the observed I/O-pair is not contained in the plant behavior, a fault different from f_i is known to be present. Then, a hypothesis different from h_i is known to be true and the Single-Fault-Assumption allows to reject h_i.

In Fig. 6.8 such a situation with the nominal behavior $\mathcal{B}(\mathcal{C}_0)$ and the plant behavior in the case of the two different faults f_1 and f_2 is shown.

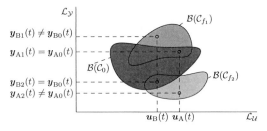

Figure 6.8: Fault hiding

If the input signal $\boldsymbol{u}_A(t)$ steers the plant into an operating region in which a fault does not have any impact on the plant behavior, the corresponding I/O-pair is contained in the nominal behavior $\mathcal{B}(\mathcal{C}_0)$. Then the fault is hidden. In Fig. 6.8 this situation occurs for fault f_1 for which the output signal $\boldsymbol{y}_{A1}(t)$ is the same as in the fault-free case, whereas fault f_2 causes another output signal. If the input signal $\boldsymbol{u}_B(t)$ is applied to the plant, fault f_2 does not have any impact on its behavior, but fault f_1 changes the plant behavior. Consecutively applying $\boldsymbol{u}_A(t)$ and $\boldsymbol{u}_B(t)$ and observing under which excitation the residual does not disappear, allows the rejection of the fault-hypothesis which corresponds to the fault which would be hidden by the input signal.

Operating regions, in which a specific fault f_i is hidden, are those operating regions, in which the edges which connect the fault variable f_i to the rest of the global structure graph are locally ineffective. A resulting directed local structure graph in such an operating region is given in Example 6.5.

Example 6.5 (Fault hiding in Plant A). *An operating region, in which fault f_4 is hidden in the behavior of Plant A is given by $\mathcal{C}_4 = \{c_{\mathrm{Elim},4}\}$ with $c_{\mathrm{Elim},4} : 0 = x_2(t)$. The corresponding directed local structure graph is depicted in Fig. 6.9. A validual $v_{4,1}(t) = y_1(t)$ can be obtained from the set $\mathcal{C}_{\mathrm{Val},4,1} = \{c_{\mathrm{Elim},4}, c_3\}$.*

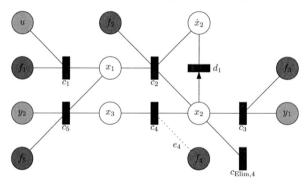

Figure 6.9: Fault hiding: directed local structure graph of Plant A

The difference of this example with respect to Example 5.3 is the following: In Example 5.3, the edge e_2 between an *unknown variable* and the rest of the local structure graph is locally ineffective. Determining the influence of this on the impact of the plant behavior in the corresponding operating region requires the analysis of the entire local structure graph. In contrast to this, in Example 6.5 the edge e_4 between a *fault variable* and the rest of the local structure graph is locally ineffective. Even without the analysis of the entire local structure graph, it is possible to conclude that the fault f_4 does not have an impact on the behavior of Plant A in the operating region described by $c_{\mathrm{Elim},4}$.

Remark 6.1. *In [105], a similar approach was taken for the distinction of faults in the input-matrix and the system-matrix of a linear system. Instead of using a structure-graph-based approach, the system matrix $\boldsymbol{A}$ is split into a nominal part $\boldsymbol{A}_0$ and a part $\boldsymbol{\Delta A}$*

which contains the faults:

$$A = A_0 + \Delta A. \tag{6.35}$$

The system is then excited with input signals $u(t)$ such that the state vector $x(t)$ lives in $\ker(\Delta A)$ *which leads to the simplification*

$$\dot{x}(t) = (A_0 + \Delta A)\, x(t) + Bu(t) = A_0 x(t) + Bu(t). \tag{6.36}$$

Then, faults in the system matrix are hidden. The input signals $u(t)$ are provided by a feedback controller which is an interesting analogy to the result in Section 6.6: Here, the input signal generator is a control law if there are input signals and output signals of the plant in the validual which was used to determine the test signal.

In this section, a special case of automatic tests was presented. It consists of an input generator that steers the plant into an operating region, in which a particular fault does not have any impact on the plant behavior. In this case the fault is said to be hidden. If the plant is in such an operating region and the current I/O-pair is inconsistent with the nominal plant behavior, the hypothesis that the hidden fault is present can be rejected. The operating region in which a fault is hidden was characterized by the conditions under which the edges which connect the fault variable with the rest of the global structure graph are locally ineffective. A parallel to a similar result for linear systems in [105] was drawn.

6.9   Summary

In this chapter, an approach to the design of automatic tests was presented. Its main idea is to identify specific operating regions in which specific local structures hold. The plant is steered into such an operating region and the I/O-pair observed at the plant is checked for consistency with the nominal plant behavior. An automatic test consists of an input generator which steers the plant into the desired operating region and a diagnostic unit which provides the refined diagnostic result.

The diagnostic unit makes use of two types of signals: the validuals which can be used to infer on the presence of an operating region, and the local residual which was introduced in Chapter 5. Dynamical systems which provide validuals can be determined with the algorithm FINDVALIDUALS on the basis of global structural analysis.

An important result is given in Theorem 6.1. It provides the set of faults which may be present if the validuals are zero, thus indicating the presence of an operating region, and the local residual is nonzero. These faults are not only the faults which affect the constraints used to determine the local residual, but the faults which affect the constraints used to determine the validuals as well.

The reason for this is that the validuals are designed under the assumption that the plant is fault-free. In this way, the difficult task of detecting an operating region in a potentially faulty plant can be avoided. Instead, the faults wich may prohibit the correct detection of the operating region are taken into account by the decision logic which provides the test result.

An approach to the design of all possible tests is given with the algorithm FINDTESTS. In this algorithm Theorem 6.1 is used to determine the hypotheses a specific test may reject.

The method for the design of input generators for the automatic tests consists of determining input signals which steer the validuals in the desired way, that is $v(t) = 0$. Solving the validual generators for the input variables, constraints on the inputs are obtained. Signals which satisfy these constraints are the test signals which steer the plant into the desired operating region.

Finally, a special type of tests which hide the impact of faults on the behavior of the plant is analyzed.

The described methods allow to determine subsets of constraints which can be used to derive an automatic test. Their main advantage is that they do not require a complete, parameterized model of the plant. The hypotheses-rejection properties of a test can be determined using structural analysis only. In general, the tests obtained in the described way allow a more precise refinement of the intermediate diagnostic result than global residual generators do. This was shown for the running example Plant A in Example 6.3.

Chapter 7

Application to a Throttle Valve

7.1 System Description

Throttle valves are actuators used in combustion engines to control the air flow to the intake manifold. A faulty throttle valve may lead to abnormal behavior of the overall engine which results in power loss and poor emission quality. A throttle valve typically consists of a DC motor driving the valve flap which is connected to a return spring. For safety-reasons, two contrariwise mounted sensors measure the position of the valve, thus providing the outputs $y_1(t)$ and $y_2(t)$. The DC motor is supplied by the voltage $U_A(t)$ which is controlled by the input $u_1(t)$. A sketch of the equivalent electro-mechanical system is depicted in Fig. 7.1.

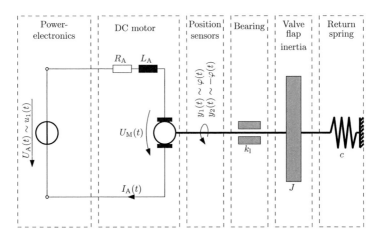

Figure 7.1: Throttle valve - electro-mechanically equivalent system

The throttle valve may be subject to the following faults:

- f_1 – offset of the supply voltage,

- f_2 – a change in the resistance R_A due to worn-out brushes,

- f_3 – a damaged or broken spring,

- f_4 – offset of the first position sensor,

- f_5 – offset of the second position sensor.

7.2   Analytical System Model

In this section, electrical and mechanical relations are used to derive an analytical model of the potentially faulty throttle valve. The modeling proceeds from the inputs of the plant to its outputs.

The power electronics which drive the system are modeled as the proportional conversion of the dimensionless control-input $u_1(t)$ to the voltage $U_A(t)$. The offset-fault of the power electronics is called f_1 and modeled by an arbitrary additional voltage to the output of the power electronics. This results in

$$U_A(t) = k_P u_1(t) + f_1 \tag{7.1}$$

which can be expressed in constraint form as

$$c_1: \quad 0 = U_A(t) - k_P u_1(t) - f_1. \tag{7.2}$$

The voltage $U_A(t)$ drives the current $I_A(t)$ through the brushes and the armature winding of the DC motor which are modeled as a resistance R_A and an inductivity L_A. The movement of the armature induces a voltage $U_M(t)$ in the armature circuit. The fault f_2 describes wear of the brushes of the DC motor which mainly results in a change of the armature resistance. The convention that in the nominal case a fault variable is zero allows to model this fault by a multiplication of the nominal armature resistance with the factor $(1 + f_2)$. Applying Kirchoff's rule to the armature circuit, one obtains

$$U_A(t) - U_M(t) = (1 + f_2) R_A I_A(t) + L_A \dot{I}_A(t) \tag{7.3}$$

which can be written in residual form as

$$c_2: \quad 0 = U_A(t) - (1 + f_2) R_A I_A(t) - L_A \dot{I}_A(t) - U_M(t). \tag{7.4}$$

The induced voltage $U_M(t)$ is proportional to the angular velocity $\dot{\varphi}(t)$ of the armature:

$$U_M(t) = k_M \dot{\varphi}(t). \tag{7.5}$$

Expressed in residual form, the constraint reads

$$c_3: \quad 0 = U_M(t) - k_M \dot{\varphi}(t). \tag{7.6}$$

The current $I_A(t)$ produces a torque $M(t)$ on the mechanical part of the throttle valve, that is the valve flap and the return spring. The torque is proportional to the current with the factor k_M:

$$M(t) = k_M I_A(t). \tag{7.7}$$

This can be written as

$$c_4: \quad 0 = M(t) - k_M I_A(t) \tag{7.8}$$

in residual form. The equation governing the mechanical part which is a mass-spring-damper system is derived from Newton's law. The torques are:

- the torque $M(t)$ generated by the DC motor,

- the torque subsequent to the acceleration $\ddot{\varphi}(t)$ of the inertia J of the valve flap,

- the torque subsequent to the angular speed $\dot{\varphi}(t)$ and the friction coefficient k_1, and

- the torque subsequent to the position $\varphi(t)$ and the spring-constant c.

Again, the multiplication of the factor $(1 + f_3)$ with the spring-constant c is used to model a faulty spring. Using this approach, a broken spring is represented by $f_3 = -1$. The torque balance then reads

$$J\ddot{\varphi}(t) = M(t) - k_1\dot{\varphi}(t) - (1 + f_3)c\varphi(t) \tag{7.9}$$

and the corresponding constraint is

$$c_5: \quad 0 = M(t) - J\ddot{\varphi}(t) - k_1\dot{\varphi}(t) - (1 + f_3)c\varphi(t). \tag{7.10}$$

Two sensors measure the angular position $\varphi(t)$ of the valve, thus providing the dimensionless outputs $y_1(t)$ and $y_2(t)$. The sensors are mounted contrariwise and can be subject to the sensor faults f_4 and f_5, respectively. The equations describing the measurements are

$$y_1(t) = k_s\varphi(t) + f_4 \tag{7.11}$$
$$y_2(t) = -k_s\varphi(t) + f_5 \tag{7.12}$$

which yields

$$c_6: \quad 0 = k_s\varphi(t) - y_1(t) - f_4 \tag{7.13}$$
$$c_7: \quad 0 = -k_s\varphi(t) - y_2(t) - f_5 \tag{7.14}$$

when expressed in constraint form. The differential relations between the variables and their derivatives w.r.t. time[1] are

$$\dot{I}_A(t) = \frac{d}{dt}I_A(t) \tag{7.15}$$

$$\dot{\varphi}(t) = \frac{d}{dt}\varphi(t) \tag{7.16}$$

$$\ddot{\varphi}(t) = \frac{d}{dt}\dot{\varphi}(t) \tag{7.17}$$

---

[1]Note that in this context the dot $(\dot{\cdot})$ does not signify the operation of differentiation but is a part of the variable name.

which yields

$$d_1 : \quad 0 = \dot{I}_A(t) - \frac{\mathrm{d}}{\mathrm{d}t} I_A(t) \tag{7.18}$$

$$d_2 : \quad 0 = \dot{\varphi}(t) - \frac{\mathrm{d}}{\mathrm{d}t} \varphi(t) \tag{7.19}$$

$$d_3 : \quad 0 = \ddot{\varphi}(t) - \frac{\mathrm{d}}{\mathrm{d}t} \dot{\varphi}(t) \tag{7.20}$$

in residual form. This completes the set of constraints $\mathcal{C}$ which describe the behavior of the throttle valve. In summary, the constraints governing the system's behavior are:

$$c_1 : \quad 0 = U_A(t) - k_P u_1(t) - f_1$$

$$c_2 : \quad 0 = U_A(t) - (1 + f_2) R_A I_A(t) - L_A \dot{I}_A(t) - U_M(t)$$

$$c_3 : \quad 0 = U_M(t) - k_M \dot{\varphi}(t)$$

$$c_4 : \quad 0 = M(t) - k_M I_A(t)$$

$$c_5 : \quad 0 = M(t) - J\ddot{\varphi}(t) - k_1 \dot{\varphi}(t) - (1 + f_3) c \varphi(t)$$

$$c_6 : \quad 0 = k_s \varphi(t) - y_1(t) - f_4$$

$$c_7 : \quad 0 = -k_s \varphi(t) - y_2(t) - f_5$$

$$d_1 : \quad 0 = \dot{I}_A(t) - \frac{\mathrm{d}}{\mathrm{d}t} I_A(t)$$

$$d_2 : \quad 0 = \dot{\varphi}(t) - \frac{\mathrm{d}}{\mathrm{d}t} \varphi(t)$$

$$d_3 : \quad 0 = \ddot{\varphi}(t) - \frac{\mathrm{d}}{\mathrm{d}t} \dot{\varphi}(t)$$

Hence, the analytical model of the throttle valve is given by

$$\mathcal{C} = \{c_1, c_2, c_3, c_4, c_5, c_6, c_7, d_1, d_2, d_3\} \tag{7.21}$$

and the variables $\mathcal{Z}$ being the union of the unknown variables $\mathcal{X}$, the known variables $\mathcal{K}$ and the fault variables $\mathcal{F}$:

$$\mathcal{X} = \{U_A, U_M, I_A, \dot{I}_A, M, \varphi, \dot{\varphi}, \ddot{\varphi}\}, \tag{7.22}$$

$$\mathcal{K} = \{u_1, y_1, y_2\}, \tag{7.23}$$

$$\mathcal{F} = \{f_1, f_2, f_3, f_4, f_5\}. \tag{7.24}$$

This section introduced the analytical model of the throttle valve. This model is used for the global structural analysis and the test generation in the next two sections.

7.3 Global Structural Analysis

In this section, the global structure of the model derived in the previous section will be analyzed. The result will serve as a benchmark for tests which can be developed by the methods described in Chapter 6.

The directed global structure graph G of the analytical model of the throttle valve is depicted in Fig. 7.2. The corresponding incidence matrix is given in Tab. 7.1.

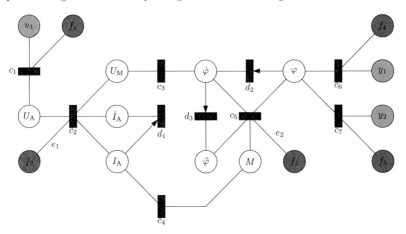

Figure 7.2: Directed global structure graph G of the throttle valve

Table 7.1: Incidence matrix M of the directed global structure graph G

M	U_A	U_M	I_A	$\dot{I}_A$	M	φ	$\dot{\varphi}$	$\ddot{\varphi}$	u_1	y_1	y_2	f_1	f_2	f_3	f_4	f_5
c_1	1								1			1				
c_2	1	1	1	1									1			
c_3		1					1									
c_4			1		1											
c_5					1	1	1	1						1		
c_6						1				1					1	
c_7						1					1					1
d_1			-1	1												
d_2						-1	1									
d_3							-1	1								

$$\underbrace{\hphantom{U_A\;U_M\;I_A\;\dot{I}_A\;M\;\varphi\;\dot{\varphi}\;\ddot{\varphi}}}_{M_{\mathcal{X}}}\quad\underbrace{\hphantom{u_1\;y_1\;y_2}}_{M_{\mathcal{K}}}\quad\underbrace{\hphantom{f_1\;f_2\;f_3\;f_4\;f_5}}_{M_{\mathcal{F}}}$$

Applying the algorithm FINDMSO to the directed global structure graph G, three GMSOs are found. Since there are causal matchings on all three GMSOs, it is possible to derive global residual generators from these GMSOs under the condition that potentially present loops can be solved. The GMSOs and the corresponding global signature matrix are given in Tab. 7.2.

Table 7.2: Global signature matrix

$\mathcal{C}_{\text{GMSO},i}$	f_1	f_2	f_3	f_4	f_5
$\{c_1, c_2, c_3, c_4, c_5, c_6, d_1, d_2, d_3\}$	1	1	1	1	
$\{c_1, c_2, c_3, c_4, c_5, c_7, d_1, d_2, d_3\}$	1	1	1		1
$\{c_6, c_7\}$				1	1

All faults are globally structurally detectable, but the faults f_1, f_2 and f_3 form a group of faults that are globally structurally not discriminable. This is the case because the fault variables f_1, f_2 or f_3 appear in the same GMSOs, c.f. Definition 4.12. Therefore, by comparing the residuals $r_i(t)$ obtained from the GMSOs in Tab. 7.2 with thresholds and using Boolean logic, it is not possible to distinguish the presence of one of these faults.

In this section, the analysis of the directed global structure graph of the throttle valve was carried out. The result is that there is a group of faults which are not globally structurally discriminable. The next section therefore aims at determining a test which will allow to distinguish between the faults in this group.

7.4   Test Generation

In this section the algorithms developed in Chapter 6 are used to determine an automatic test for the throttle valve. This is motivated by the observation that global residual generators which are obtained from the GMSOs found in Section 7.3 do not allow to distinguish the faults f_1, f_2 and f_3.

The algorithm FINDTESTS typically finds a large number of possible tests, the sensitivity of which can be determined by the help of the local and directed global structure graphs. Its steps are carried out for the throttle valve in the following.

Applying the algorithm FINDTESTS yields two edges which are locally ineffective in the first step. The edges and the corresponding operating regions are

$$e_1 \text{ (between } c_2 \text{ and } f_2) \text{ which is locally ineffective under } \quad c_{\text{Elim},1} : \ 0 = I_{\text{A}}(t) \quad (7.25)$$

and

$$e_2 \text{ (between } c_5 \text{ and } f_3) \text{ which is locally ineffective under } \quad c_{\text{Elim},2} : \ 0 = \varphi(t). \quad (7.26)$$

The sets $\mathcal{C}_{\text{Elim}}$ and $\mathcal{E}_{\text{Elim}}$ are therefore

$$\mathcal{E}_{\text{Elim}} = \{e_1, e_2\} \quad \text{and} \quad \mathcal{C}_{\text{Elim}} = \{c_{\text{Elim},1}, c_{\text{Elim},2}\}. \quad (7.27)$$

The second step reveals that all operating region candidates are valid operating regions and, therefore,

$$\mathcal{A}_{\text{VOR}} = 2^{\mathcal{C}_{\text{Elim}}} = \{\{\emptyset\}, \{c_{\text{Elim},1}\}, \{c_{\text{Elim},2}\}, \{c_{\text{Elim},1}, c_{\text{Elim},2}\}\} \tag{7.28}$$

holds. This also implies that all corresponding local structure graphs are feasible. In Steps 3-7 a large number of tests is found. This is due to the large number of permutations of possible validuals. In Step 8, Theorem 6.1 is used to determine the sensitivity of these tests. In order to distinguish whether fault f_1 or f_2 is present, the following test is chosen in Step 5 of the algorithm:

$$\text{Local residual:} \qquad \mathcal{C}_{\text{Res},1,1} = \{c_1, c_2, c_3, c_6, d_1, d_2, c_{\text{Elim},1}\} \tag{7.29}$$

$$\text{Validual:} \qquad \mathcal{C}_{\text{Val},1,1} = \{c_4, c_5, c_6, d_2, d_3, c_{\text{Elim},1}\} \tag{7.30}$$

In the following, the sensitivity analysis which is carried out in Step 8 of the algorithm FINDTESTS is done for this test. If the throttle valve is in the operating region defined by $c_{\text{Elim},1}$, the local structure graph

$$G|\{c_{\text{Elim},1}\} = ((\mathcal{C} \cup c_{\text{Elim},1}) \cup \mathcal{Z}, \mathcal{E} \setminus e_1) \tag{7.31}$$

holds. Computing the sensitivity of the test with the help of Theorem 6.1 and the global and the local structure graphs yields that a test based on the LMSO $\mathcal{C}_{\text{Res},1,1}$ and the GMSO $\mathcal{C}_{\text{Val},1,1}$ is sensitive to the faults f_1, f_3 and f_4 because

$$\mathcal{F}_{\text{Con}} = \mathcal{F} \cap \text{var}^{G(\mathcal{C})|c_{\text{Elim},1}} (\mathcal{C}_{\text{Res},1,1}) = \{f_1, f_4\} \tag{7.32}$$

and

$$\mathcal{F}_{\text{Ste}} = \mathcal{F} \cap \text{var}^{G(\mathcal{C} \cup c_{\text{Elim},1})} (\mathcal{C}_{\text{Val},1,1}) = \{f_3, f_4\} \tag{7.33}$$

hold. Note that eqn. (7.32) does not contain f_2 because it was determined using the local structure graph in which the edge e_1 depicted in the global structure graph in Fig. 7.2 is locally ineffective. Therefore the fault variable f_2 is not connected with the rest of the local structure graph. It is here, where the fault-effect-chain between the fault and the behavior is interrupted. With the results in Section 6.5, it is possible to conclude that if the local residual is nonzero and the validual is zero, it is possible to reject the hypotheses

$$\mathcal{H}_{\text{Rej}} = \{h_2, h_5\}. \tag{7.34}$$

Note also that these results are obtained with by help of the structure graphs only. It was not necessary to determine the analytical form of the residuals and validuals in order to determine the sensitivity of the test. This is a major advantage of the design method for automatic tests developed in this thesis.

Eliminating the unknown variables from the LMSO $\mathcal{C}_{\text{Res},1,1}$, one obtains the local residual generator

$$r_1(t) = k_{\text{P}} u_1(t) - \frac{k_{\text{M}}}{k_{\text{S}}} \frac{\text{d}}{\text{d}t} y_1(t). \tag{7.35}$$

Solving the GMSO $\mathcal{C}_{\mathrm{Val},1,1}$ for the only unknown variable $I_{\mathrm{A}}(t)$ in the constraint $c_{\mathrm{Elim},1}$ representing the operating region, and injecting the result into the constraint $c_{\mathrm{Elim},1}$, the validual generator

$$v_1(t) = \frac{1}{k_{\mathrm{M}} k_{\mathrm{S}}} \left(J \frac{\mathrm{d}^2}{\mathrm{d}t^2} y_1(t) + k_1 \frac{\mathrm{d}}{\mathrm{d}t} y_1(t) + c y_1(t) \right) \tag{7.36}$$

is obtained. The test signal generator can be realized by a controller which actively steers the validual $v_1(t)$ to zero using the input signal $u_1(t)$.

The physical explanation of this test is as follows: If the validual $v_1(t)$ is zero, the power electronics are actuated in a way that the torque $M(t)$ becomes zero. Since for a DC motor, the torque is directly proportional to the current $I_{\mathrm{A}}(t)$ it is then likewise zero. If the current $I_{\mathrm{A}}(t)$ is constantly zero, its derivative $\dot{I}_{\mathrm{A}}$ is zero as well. Then, there is no voltage drop over the inductivity L_{A} and the resistance R_{A} which is influenced by the fault f_2 – independently of the value of the fault. It is here, where the fault effect-chain is interrupted.

If the current $I_{\mathrm{A}}(t)$ and its derivative w.r.t. time $\dot{I}_{\mathrm{A}}(t)$ are zero, the induced voltage $U_{\mathrm{M}}(t)$ equals the voltage $U_{\mathrm{A}}(t)$ which is controlled by the input $u_1(t)$. This explains why the local residual $r_1(t)$ is basically a comparison of these two voltages.

The above reasoning also explains, why the validual is the torque balance of the unactuated valve. If this torque balance holds true, that is $v_1(t) = 0$, the torque $M(t)$ is zero.

Concluding on the presence of the operating region $c_{\mathrm{Elim},1}$ with the validual $v_1(t)$ may fail due to a fault in the position sensor (f_4) and a broken or damaged spring (f_3), a result which was already found using structural analysis in eqn. (7.32). If the throttle valve is in the specific operating region defined by $I_{\mathrm{A}}(t) = 0$, the only faults which may have an impact on the comparison of the voltages which is realized by the local residual $r_1(t)$ are a fault in the power electronics (f_1) and a faulty position sensor (f_4). The fault f_2 which is a change in the armature resistance does not have an impact on the comparison of the voltages: since the current through the armature resistance is zero, there is no voltage drop over R_{A} – independently of the value of R_{A}. Then, a fault which changes the armature resistance does not have any impact on the voltages in the armature circuit.

This result has also been obtained by the analysis of the local structure graph, see eqn. (7.33). Hence, the faults to which the overall test is sensitive are f_1, f_3 and f_4. This allows to distinguish the fault f_2 from the faults f_1 and f_3, which is not possible using global residual generators only.

The above explanation shows that automatic tests can also be found by physical reasoning. However, the method presented in this thesis replaces this rather complicated approach and allows to find all possible tests.

In this section, the algorithm FINDTESTS was used to determine a test for the throttle valve. This test allows to discriminate faults which are not globally structurally discriminable. The test is proven to be successful in a simulation study in the next section.

7.5   Simulation Study

In this section, the automatic test for the throttle valve which was obtained in the previous section is analyzed in a simulation study.

Situation.   In the following it is assumed that the intermediate diagnostic result was already refined to the hypothesis-set $\mathcal{H}(1) = \{h_1, h_2, h_3\}$ by the global structural approaches. Residual generators found with global structural approaches do not allow a further reduction of this set of fault-hypotheses. In order to refine this intermediate result, the test found in Section 7.4 is applied to the throttle valve. In the following the application of this test is simulated for the fault-free case and the three fault scenarios f_1, f_2 and f_3. In this example not only the local residual and the validual are dimensionless quantities but the inputs and outputs as well. Therefore, no units are given in the graphs.

Simulation of the test.   In Fig. 7.3, the simulation result of the test applied to the fault-free throttle valve is given. Until time $t = 23$s, the plant is excited with a sine wave. Although there is no fault in the throttle valve, the local residual is not zero. This is because the corresponding operating region is not reached which is indicated by the nonzero validual. It is this effect that makes the difference between the local and the global residual: Whereas the global residual needs to be zero in the fault-free case for arbitrary signal inputs, the local residual is zero in the fault-free case only under excitation with specific input signals. For this test, these specific input signals are obtained with a PI-controller which makes the validual zero. At $t = 23$s it is switched from the sinusoidal excitation to the excitation with the controller. The validual $v_1(t)$ then becomes zero at $t = 45$s indicating that the operating region was reached. Since the throttle valve is considered to be fault-free in this simulation, the local residual $r_1(t)$ becomes zero as well.

In Fig. 7.4, the simulation results of the test in the three fault scenarios f_1, f_2 and f_3 are given: Similar to the simulation of the fault-free plant, the test is applied to the valve at $t = 23$s and the validual $v_1(t)$ becomes zero at $t = 45$s. Once the latter is the case, the local residual can be used for fault diagnosis. In the following, each fault-scenario is considered separately for this instant of time.

Test result for scenario f_1.   If the fault f_1, that is an offset of the supply voltage is present, the test results in the validual $v_1(t)$ being zero and the local residual $r_1(t)$ being nonzero. The observed I/O-pair is therefore inconsistent with the behavior of the throttle valve in the operating region $c_{\mathrm{Elim},1}$. Since the set of hypotheses $\mathcal{H}_{\mathrm{Rej}}$ that can be rejected is known from the structural analysis during the design of the test with eqn. (6.20) the intermediate diagnostic result $\mathcal{H}(1)$ is refined to

$$\mathcal{H}(2) = \mathcal{H}(1) \setminus \mathcal{H}_{\mathrm{Rej}} = \{h_1, h_2, h_3\} \setminus \{h_2, h_4\} = \{h_1, h_3\}. \tag{7.37}$$

Note that in this case from $v_1(t) = 0$ it is possible to conclude that the valve is actually in the operating region defined by $c_{\mathrm{Elim},1}$ and the directed local structure graph $G|c_{\mathrm{Elim},1}$ which was used to determine the local residual generator $r_1(\boldsymbol{u}(t), \boldsymbol{y}(t), \frac{\mathrm{d}}{\mathrm{d}t})$ holds. The reason for this is that the fault variable f_1 does not appear in the constraint set which is used to determine the validual, c.f. eqn. (7.33).

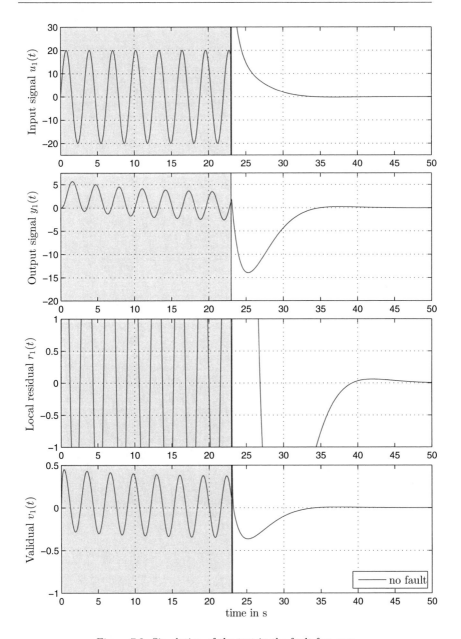

Figure 7.3: Simulation of the test in the fault-free case

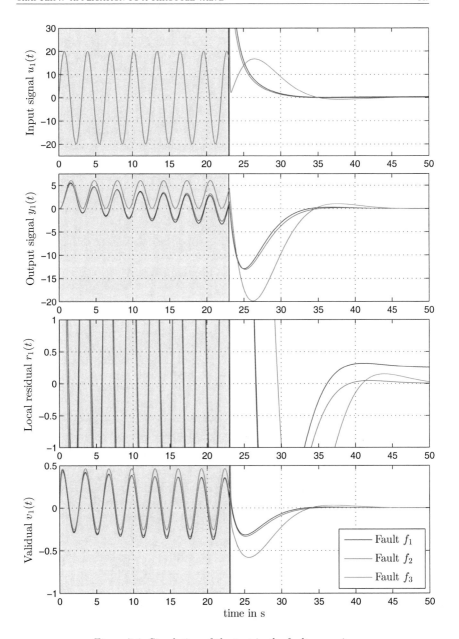

Figure 7.4: Simulation of the test in the fault scenarios

Test result for scenario f_2. If the fault f_2, that is wear of the brushes, is present in the valve, the test results in the validual $v_1(t)$ being zero and the local residual $r_1(t)$ being zero as well. No inconsistency between the observed I/O-pair and the I/O-behavior of the valve in this operating region can be detected. The reason for this is that the test is not sensitive to fault f_2:

$$\mathcal{F}_{\text{Sens}} = \{f_1, f_3, f_4\}. \tag{7.38}$$

This result was obtained by the structural analysis which was conducted during the design of the test. Therefore, if the fault which is actually present in the throttle valve is f_2, the test is unsuccessful. The result of the test is

$$\mathcal{H}(2) = \mathcal{H}(1) = \{h_1, h_2, h_3\}, \tag{7.39}$$

the intermediate diagnostic result is not refined.

Test result for scenario f_3. In the case of fault f_3, that is the spring in the valve is broken, the test results in the validual $v_1(t)$ being zero, and the local residual $r_1(t)$ being nonzero. The test refines the diagnostic result:

$$\mathcal{H}(2) = \mathcal{H}(1) \setminus \mathcal{H}_{\text{Rej}} = \{h_1, h_2, h_3\} \setminus \{h_2, h_5\} = \{h_1, h_3\}. \tag{7.40}$$

Note that this is the interesting case, in which the fault variable does not appear in the constraint set which was used to derive the local residual, but in the constraint set which was used to determine the validual. It is therefore $\mathcal{F}_{\text{Ste}}$ in eqn. (6.14). Although the validual was zero during the test, the operating region which corresponds to the directed local structure graph used to determine the local residual generator $r_1(\boldsymbol{u}, \boldsymbol{y}, \frac{\mathrm{d}}{\mathrm{d}t})$ was not reached. Nevertheless, the rejection of a wrong fault-hypothesis was possible. This shows the convenience of the solution suggested for the difficult problem of steering a potentially faulty system.

Summary of the simulation. The simulation of the test for the throttle valve obtained in the previous section was done for the fault-free plant and the fault scenarios f_1, f_2 and f_3. Whereas these faults are not discriminable using global residuals without dedicated excitation, they can be distinguished with the test signal applied here. Once the test signal is applied to the system, the validual $v_1(t)$ becomes zero. Then, the value of the local residual is used to determine whether the fault-hypotheses $\mathcal{H}_{\text{Rej}}$ which were found using structural analysis during the design of the test can be rejected. In that way, the discrimination of faults which are not globally structurally discriminable becomes possible.

7.6   Summary

In this chapter, the method to determine automatic tests developed in the previous chapters of this thesis was exemplified with a throttle valve. This typical automotive system can be subject to a number of faults. The throttle valve was described in Section 7.1 and modeled in Section 7.2. The global structural analysis of this model in Section 7.3

resulted in three GMSOs, from which global residual generators can be obtained. A main result of the analysis of the directed global structure graph was that a group of the faults cannot be distinguished by the global residual generators, threshold checking and a Boolean logic.

In Section 7.4, the method to design automatic tests from Chapter 6 was applied to the throttle valve. The method resulted in an input generator, a local residual, a validual and a decision logic forming an automatic test. This test allowed to distinguish faults which were not discriminable using the global structural approach. A physical explanation of the test found with the methodical approach based on the local structural analysis was given. The functional capability of the test was shown in simulation in Section 7.5.

Chapter 8

Conclusions and Outlook

Conclusion

The goal of service diagnosis is the identification of the faulty component in a defective system. This problem is solved by using tests rejecting wrong fault-hypotheses. The consecutive execution of tests results in a sequential process in which the intermediate diagnostic result is refined stepwise. Eventually, this leads to the fault that is actually present in the system.

In this thesis, a model-based method for the construction of automatic tests is developed. The resulting tests use the dedicated excitation of the system to be diagnosed in combination with a consistency test of the observed input and output signals. An inconsistency of these signals and the behavior of the fault-free system allows the rejection of wrong fault-hypotheses. The automatic tests found with this method can therefore be used in the sequential process of service diagnosis.

The method developed in this thesis allows to determine the test signals as well as a specific way to realize a consistency test. This consistency test considers the behavior of the fault-free system in a specific operating region. In order to account for the specific nature of mechatronic systems (dynamical, nonlinear, high complexity), the approach taken in this thesis is based on an abstraction of the system model. This abstraction is a structure graph describing the couplings in the system. This graph takes into account that the system is potentially faulty and in a specific operating region.

The approach is derived from an analysis of the properties a potentially faulty system needs to satisfy in order to allow the detection and distinction of faults on the basis of the input and output signals of the system.

Determining the test signals. The method of determining test signals for fault diagnosis consists of three steps:

1. Specific operating regions are identified. This is done with the help of the *global structure graph* of the system. In this graph, the couplings between variables and constraints are represented by edges in a bipartite graph. The method to determine the operating regions consists of investigating for each edge under which additional constraints on the variables the coupling represented by the edge vanishes. An operating region is then defined by a set of such additional constraints. The first main contribution of this thesis is the *local structure graph* which is the graph

151

that describes the couplings between the variables and the constraints if such an operating region is present. The structural difference between the local and the global structure graph is that in the local structure graph the edges representing the vanished couplings are not present and additional constraints representing the operating region are included.

2. *Validuals*, signals that allow to infer on the presence of the operating region and, therefore, the validity of the local structure graph, are determined. This new kind of signal is the second main contribution of this thesis. A validual is calculated from the system's input and output signals and indicates the presence of the operating region in the fault-free system if it is zero. The way it is computed is determined by the help of a specific global structure graph: namely the directed global structure graph of the constraints which govern the system-behavior and the constraint which corresponds to an operating region in which an edge in the global structure graph vanishes.

3. The test signals are found by determining which signals steer the system in a way such that the validuals become zero. Depending on the nature of the validuals, this approach results in a feedforward controller, a control-law, or a feedback controller for the generation of the test signals.

Evaluation of the input and output signals. In the automatic tests a specific kind of consistency test is used to reject wrong fault-hypotheses. The method of designing this consistency test consists of two steps:

1. *Local residuals* which are an extension of *global residuals* (known from the literature simply as residuals) are designed. This new kind of signals realizes a consistency test of the input and output signals and the nominal behavior under the assumption that the system is in a specific operating region. Whereas global residuals check the consistency in an arbitrary operating region and, therefore, need to be zero in the fault-free case for arbitrary input signals, the definition of the new local residuals relaxes this restrictive property: Local residuals only need to be zero in the fault-free case for specific input signals. The local residual generators are found by applying algorithms to determine minimal overconstrained constraint sets to the local structure graph which is present if the corresponding validuals are zero. Then the unknown variables are eliminated from a constraint set found in that way. This procedure results in a local residual. Generally, more local residuals than global residuals can be found indicating the better fault discrimination properties of these signals and assuming that the system has been brought to the corresponding operating region.

2. Finally, a Boolean logic is designed which allows to reject wrong fault-hypotheses on the basis of the comparison of a local residual and the validuals with thresholds. This integration of the validuals, a local residual and the decision logic to the diagnostic unit of an automatic test is the third main contribution of this thesis. The design of the decision logic is based on the analysis of the structure graphs

used during the design of the validuals and the local residual. The approach takes
into account that steering the faulty system into an operating region may actually
fail. This comprises the interesting case that an operating region may actually
not be reached although the validual is zero and thus indicates the presence of the
operating region. This situation may occur because a fault may have an impact on
the part of the system model which was used to calculate the validual. Taking into
account this effect in the decision logic, the difficult problem of steering a system
potentially subject to a fault is solved.

The methods were used to determine an automatic test for a throttle valve which
is a typical automotive system. It was shown that this test allows to distinguish faults
which cannot be distinguished using global structural analysis and the resulting global
residual generators. This is reached by the dedicated system excitation in the test. The
test obtained with the new method results in a physically interpretable way to distinguish
the faults.

A number of publications has led to the integrated approach described in Chapter 6
of this thesis. The original reasoning started from the idea to eliminate the couplings
represented by the edges between fault variables and the rest of the global structure graph.
This was introduced with the example of a spark ignition engine in [2] and detailed by
using the example of a throttle valve in [5]. The generalization to the local structure
graph and the validation of the corresponding operating region by validuals as well as
their integration in automatic tests was introduced in [3] and detailed in [6]. In [4], an
algortihm was described to determine all possible tests.

The approach described in this thesis was successfully applied to a number of automotive systems which led to the patent applications [1], [8], [9] and [10].

The methods presented in this thesis are particularly suited for nonlinear and large-
scale systems. This is not only due to the structural approach to the problem which
has earlier been shown to be advantageous for this type of systems. It is also a result
of the idea to identify operating regions in which couplings vanish. Only one coupling
represented by an edge between a variable and a constraint in the global structure graph
is considered at a time. This is considerably simpler than analyzing the entire state space
representation of a system in order to obtain states in which a fault does not influence
the system's behavior.

A further advantage is the partial automation of the approaches. This is possible
because the determination of the operating regions and the constraint sets, from which
local residuals and validuals can be obtained, are based on structural representations.
This allows an easy use of the methods facilitating the development of automatic tests
for service diagnosis.

However, since the method presented in this thesis is based on a model of the system,
the quality of the resulting automatic tests depends on the quality of the models used.
Nowadays, the development of mechatronic systems relies on the simulation of the systems
and, therefore, the necessary models are often readily available.

Outlook

The computational burden of the algorithm to determine all tests is high. This problem, due to the large number of permutations which are considered when computing all tests, could be approached with the methods in [39]. Besides this aspect of implementation, future works should concentrate on three main topics:

Threshold design for validual evaluation. A first measure in treating robustness issues in diagnosis usually consists of filtering the residuals, implementing debouncing measures and increasing the thresholds. Generally, the rough rule of thumb applies that the larger the thresholds, the less frequent are false alarms. However, the theory developed in this thesis is based on the idea that the system is in the operating region, where the coupling represented by an edge in the global structure graph vanishes completely. This corresponds to the validual being zero. In reality, the validual is not checked to be zero, but compared to a threshold in order to infer that the operating region is actually present. Therefore, if the validual is not exactly zero, the coupling does not vanish completely. A local residual may then be nonzero, although no fault is present in the system. This can lead to a wrong test result. It is straightforward to assume that the larger one allows the threshold for the validual to be, the larger the threshold for the local residual needs to be chosen. However, a detailed analysis of this subject is needed in order to ensure that the tests do not reject true fault-hypotheses. One approach to this problem could be to transform the threshold on the validual to disturbances of the variables occurring in the description of the operating region. The impact of this disturbance could then be analyzed and the threshold for the local residual could be chosen accordingly.

Hierarchical approach to validual generation. The method to determine validuals developed in this thesis is based on the global structure graph of the model-constraints and the constraint under which an edge is locally ineffective. This approach is also pursued if more than one edge is eliminated from the global structure graph. However, if one validual is zero, the corresponding local structure graph already holds. For this reason, it is possible to determine the second validual generator with the help of the local structure graph which holds if the first validual is zero. The third validual can then be found using the local structure graph which results from the elimination of the first two edges and so on. The signals found in that way only have the property of a validual if the previous validuals are zero. They are therefore *local validuals*. Since generally the degree of structural redundancy is larger in local structure graphs, this approach results in a larger set of possible validual generators, the benefit of which is better fault discrimination. However, determining which faults a test resulting from a local residual and local validuals is sensitive to, is difficult. The result will be a more complex relationship than the sensitivity relation in Theorem 6.1. The extension of this relation to the case of local validuals would be a first step towards their application.

Integration in automatic design. In order to further facilitate the application of automatic tests, the method presented in this thesis should be integrated in a design tool. Some efforts towards this direction have already been made in the course of this thesis, c.f. [73]: namely a toolbox which allows to determine locally ineffective edges and the constraint sets from which validuals and local residuals can be computed.

However, determining validual generators and local residual generators still requires manual manipulation of the constraints. Symbolic solvers and methods known from the numerical solution of differential equations could be applied to automatically generate the validual and local residual generators from the constraint sets. For this purpose, the recent results in [116] for the automated design of FDI-systems could be used in the context of test design.

Bibliography

Articles and Patent Applications by the Author

[1] Michael Ungermann, Andreas Buse, and Dieter Schwarzmann. Verfahren und Vorrichtung zum Erkennen von Lecks im Einlasstrakt eines Verbrennungsmotors. German patent application, DE102010029021 A1, 2011. Also published in:

- Procedimento e dispositivo per riconoscere fughe per difetto di tenuta nel trarro di ammissione di un motore endotermico, Italian patent application, ITMI20110697 A1

- Method and device for detecting leaks in the intake tract of an internal combustion engine, PR China patent application, CN102251865 A

- Method and device for detecting leaks in the intake tract of an internal combustion engine, US patent application, US20110282598 A1

[2] Michael Ungermann, Jan Lunze, and Dieter Schwarzmann. Model-based test signal generation for service diagnosis of automotive systems. In *Proceedings of the 6th IFAC Symposium on Advances in Automotive Control (AAC'10)*, pages 117–122, Munich, Germany, 2010.

[3] Michael Ungermann, Jan Lunze, and Dieter Schwarzmann. Service diagnosis utilizing the dependencies between the system structure and the operating points. In *Proceedings of the Conference on Control and Fault-Tolerant Systems (SysTol'10)*, pages 873–878, Nice, France, 2010.

[4] Michael Ungermann, Jan Lunze, and Dieter Schwarzmann. Service diagnosis of dynamical systems based on local structural models. In *Proceedings of the 19th Mediterranean Conference on Control and Automation (MED'11)*, pages 455–460, Corfu, Greece, 2011.

[5] Michael Ungermann, Jan Lunze, and Dieter Schwarzmann. Testgenerierung für die Werkstattdiagnose auf der Basis struktureller Modelle. *at - Automatisierungstechnik*, 59(5):270–278, 2011.

[6] Michael Ungermann, Jan Lunze, and Dieter Schwarzmann. Test signal generation for service diagnosis based on local structural properties. *International Journal of Applied Mathematics and Computer Science*, 22(1):55–65, 2012.

[7] Michael Ungermann and David Schmitz. Schnell schaltendes Ventil mit Überwachungseinrichtung. German patent application, DE102011118651 A1, 2013.

[8] Michael Ungermann and Dieter Schwarzmann. Verfahren und Vorrichtung zum Erkennen einer Fehlerart in einem Stellgeber. German patent application, DE102011075047 A1, 2012.

[9] Michael Ungermann and Dieter Schwarzmann. Verfahren und Vorrichtung zum Erkennen von Klappenfehlern in einem Verbrennungsmotor. German patent application, DE102011075744 A1, 2012.

[10] Michael Ungermann and Dieter Schwarzmann. Motorsteuerung für Verbrennungsmotor. German patent application, DE102011078891 A1, 2013. Also published in:

- Engine control for internal combustion engine, International Patent Cooperation Treaty application, WO2013/007465 A1

General References

[11] Jochen Aßfalg and Frank Allgöwer. Fault diagnosis of constrained nonlinear systems using structured augmented state models. In *Proceedings of the 6th IFAC Symposium on Fault Detection, Supervision and Safety of Technical Processes (SAFEPROCESS'06)*, pages 1300–1305, Beijing, China, 2006.

[12] Jochen Aßfalg and Frank Allgöwer. Fault diagnosis with structured augmented state models: Modeling, analysis and design. In *Proceedings of the 45th IEEE Conference on Decision and Control (CDC'06)*, pages 1165–1170, San Diego, USA, 2006.

[13] Ivan V. Andjelkovic. *Auxiliary Signal Design for Fault Detection for Nonlinear Systems: Direct Approach*. PhD thesis, North Carolina State University, 2008.

[14] Ivan V. Andjelkovic, Kelly Sweetingham, and Stephen L. Campbell. Active fault detection in nonlinear systems using auxiliary signals. In *Proceedings of the 2008 American Control Conference (ACC'08)*, pages 2142–2147, Seattle, USA, 2008.

[15] Joaquim Armengol, Anibal Bregon, Teresa Escobet, Escobet Gelso, Mattias Krysander, Mattias Nyberg, Xavier Olive, Belarmino Pulido, and Louise Travé-Massuyès. Minimal structurally overdetermined sets for residual generation: A comparison of alternative approaches. In *Proceedings of the 7th IFAC Symposium on Fault Detection, Supervision and Safety of Technical Processes (SAFEPRO-CESS'09)*, pages 1480–1485, Barcelone, Spain, 2009.

[16] Alireza Esna Ashari, Ramine Nikoukhah, and Stephen L. Campbell. Auxiliary signal design for robust active fault detection of linear discrete-time systems. *Automatica*, 47(9):1887–1895, 2011.

[17] Jan Åslund, Anibal Bregon, Mattias Krysander, Erik Frisk, Belarmino Pulido, and Gautam Biswas. Structural diagnosability analysis of dynamic models. In *Proceedings of the 18th IFAC World Congress*, pages 4082–4088, Milano, Italy, 2011.

[18] Jan Åslund and Erik Frisk. Structural analysis for fault diagnosis of models with constraints. In *Proceedings of the 7th IFAC Symposium on Fault Detection, Supervision and Safety of Technical Processes (SAFEPROCESS'09)*, pages 384–389, Barcelone, Spain, 2009.

[19] Francois Bateman, Hassan Noura, and Mustapha Ouladsine. Active fault detection and isolation strategy for an unmanned aerial vehicle with redundant flight control surfaces. In *Proceedings of the 16th Mediterranean Conference on Control and Automation (MED'08)*, pages 1246–1251, Ajaccio, France, 2008.

[20] Bernd Bertsche. *Reliability in Automotive and Mechanical Engineering*. Springer, 2008.

[21] Bernard Bäker and Andreas Unger. *Diagnose in mechatronischen Fahrzeugsystemen V.* Expert Verlag, 2009.

[22] Mogens Blanke, Michel Kinnært, Jan Lunze, and Marcel Staroswiecki. *Diagnosis and Fault-Tolerant Control.* Springer, 2006.

[23] Mogens Blanke and Torsten Lorentzen. Satool - a software tool for structural analysis of complex automation systems. In *Proceedings of the 6th IFAC Symposium on Fault Detection, Supervision and Safety of Technical Processes (SAFEPROCESS'06)*, pages 673–678, Beijing, PR China, 2006.

[24] Mogens Blanke and Marcel Staroswiecki. Structural design of systems with safe behavior under single and multiple faults. In *Proceedings of the 6th IFAC Symposium on Fault Detection, Supervision and Safety of Technical Processes (SAFEPRO-CESS'06)*, pages 511–516, Beijing, PR China, 2006.

[25] Ilja N. Bronstein, Konstantin A. Semendjajew, and Gerhard Musiol. *Taschenbuch der Mathematik.* Verlag Harri Deutsch, 2000.

[26] Stephen L. Campbell, Kimberly J. Drake, Ivan V. Andjelkovic, Kelly Sweetingham, and Dongkyoung Choe. Model based failure detection using test signals from linearizations: a case study. In *Proceedings of the 2006 IEEE Conference on Computer Aided Control Systems Design (CACSD)*, pages 2659–2664, Munich, Germany, 2006.

[27] Stephen L. Campbell, Kimberly J. Drake, and Ramine Nikoukhah. Auxiliary signal design for multi-model identification in systems with multiple delays. In *Proceedings of the 10th Mediterranean Conference on Control and Automation (MED'02)*, Lisbon, Portugal, 2002.

[28] Stephen L. Campbell, Kimberly J. Drake, and Ramine Nikoukhah. Analysis of spline based auxiliary signal design for failure detection in delay systems. In *Proceedings of the 2003 IEEE International Conference on Systems, Man and Cybernetics (SMC'03)*, pages 2551–2556, Washington, D.C., USA, 2003.

[29] Steven L. Campbell and Ramine Nikoukhah. *Auxiliary Signal Design for Failure Detection*. Princeton University Press, 2004.

[30] Dongkyoung Choe. *Digital Signal Design for Fault Detection in Linear Continuous Dynamical Systems*. PhD thesis, North Carolina State University, 2007.

[31] Dongkyoung Choe, Stephen L. Campbell, and Ramine Nikoukhah. Auxiliary signal design for robust failure detection: a case study. In *Proceedings of the 5th International Conference on Control and Automation (ICCA'05)*, pages 1008–1013, Budapest, Hungary, 2005.

[32] Dongkyoung Choe, Stephen L. Campbell, and Ramine Nikoukhah. A comparison of optimal and suboptimal auxiliary signal design approaches for robust failure detection. In *Proceedings of the 2005 IEEE International Conference on Control Applications (CCA'05)*, pages 1473–1478, Toronto, Canada, 2005.

[33] Dongkyoung Choe, Stephen L. Campbell, and Ramine Nikoukhah. Optimal piecewise-constant signal design for active fault detection. *International Journal of Control*, 82(1):130–146, 2009.

[34] Vincent Cocquempot, Roozbeh Izadi-Zamanabadi, Marcel Staroswiecki, and Mogens Blanke. Residual generation for the ship benchmark using structural approach. In *Proceedings of the UKACC International Conference on Control (CONTROL'98)*, pages 1480–1485, Swansea, United Kingdom, 1998.

[35] Marie-Odile Cordier, Philippe Dague, M. Dumas, Francois Lévy, Jacky Montmain, Marcel Staroswiecki, and Louise Travé-Massuyès. A comparative analysis of ai and control theory approaches to model-based diagnosis. In *Proceedings of the 14th European Conference on Artificial Intelligence (ECAI'00)*, pages 136–140, Berlin, Germany, 2000.

[36] Marie-Odile Cordier, Philippe Dague, Francois. Levy, Jacky Montmain, Marcel Staroswiecki, and Louise Travé-Massuyès. Conflicts versus analytical redundancy relations: a comparative analysis of the model based diagnosis approach from the artificial intelligence and automatic control perspectives. *IEEE Transactions on Systems, Man, and Cybernetics, Part B: Cybernetics*, 34(5):2163–2177, 2004.

[37] Dilek Düştegör, Vincent Cocquempot, and Marcel Staroswiecki. Structural analysis for residual general: Implementation issues considerations. In *Proceedings of the International Conference on Control (UKACC'04)*, Bath, United Kingdom, 2004.

[38] Dilek Düştegör, Vincent Cocquempot, and Marcel Staroswiecki. Structural analysis for residual generation: Towards implementation. In *Proceedings of the 2004 IEEE*

International Conference on Control Applications (CCA'04), pages 1217–1222, Taipei, Taiwan, 2004.

[39] Dilek Düştegör, Vincent Cocquempot, and Marcel Staroswiecki. Adaptive structural analysis for FDI design in evolving systems. In *Proceedings of the 6th IFAC Symposium on Fault Detection, Supervision and Safety of Technical Processes (SAFEPROCESS'06)*, pages 420–425, Beijing, China, 2006.

[40] Dilek Düştegör, Erik Frisk, Vincent Cocquempot, Mattias Krysander, and Marcel Staroswiecki. Structural analysis of fault isolability in the DAMADICS benchmark. *Control Engineering Practice*, 14:597–608, 2006.

[41] Vincent de Flaugergues, Vincent Cocquempot, Mireille Bayart, and Marco Pengov. Structural analysis for FDI: a modified, invertibility-based canonical decomposition. In *Proceedings ot the 20th International Workshop on the Principles of Diagnosis (DX'09)*, pages 59–66, Stockholm, Sweden, 2009.

[42] Vincent de Flaugergues, Vincent Cocquempot, Mireille Bayart, and Marco Pengov. On non-invertibilities for structural analysis. In *Proceedings ot the 21st International Workshop on the Principles of Diagnosis (DX'10)*, Portland, USA, 2010.

[43] Fanny Djoutsop. Modeling and Active Diagnosis of a Faulty Spark Ignition Engine. Diploma Thesis, TU Darmstadt, 2010.

[44] Kimberly J. Drake. *Analysis of Numerical Methods for Fault Detection and Model Identification in Linear Systems with Delays*. PhD thesis, North Carolina State University, 2003.

[45] Kimberly J. Drake, Steven L. Campbell, Ivan V. Andjelkovic, and Kelly Sweetingham. Model-based failure detection on nonlinear systems: Theory and transition. *Naval Engineers Journal*, 119:93–107, 2007.

[46] A. L. Dulmage and N. S. Mendelsohn. Coverings of bipartite graphs. *Canadian Journal of Mathematics*, 10:516–534, 1958.

[47] Alireza Esna Ashari, Ramine Nikoukhah, and Steve Campbell. Active robust fault detection of closed-loop systems: General cost case. In *Proceedings of the 7th IFAC Symposium on Fault Detection, Supervision and Safety of Technical Processes (SAFEPROCESS'09)*, pages 585–590, Barcelone, Spain, 2009.

[48] Martene Fair and Stephen L Campbell. Active incipient fault detection with two simultaneous faults. In *Proceedings of the 7th IFAC Symposium on Fault Detection, Supervision and Safety of Technical Processes (SAFEPROCESS'09)*, pages 573–578, Barcelone, Spain, 2009.

[49] Martene Fair and Steven L. Campbell. Active incipient fault detection with more than two simultaneous faults. In *Proceedings of the 2009 IEEE International Conference on Systems, Man and Cybernetics (SMC'09)*, pages 3322–3327, San Antonio, USA, 2009.

[50] Erik Frisk, Anibal Bregon, Jan Åslund, Mattias Krysander, Belarmino Pulido, and Gautam Biswas. Diagnosability analysis considering causal interpretations for differential constraints. In *Proceedings ot the 21st International Workshop on the Principles of Diagnosis (DX'10)*, Portland, USA, 2010.

[51] Erik Frisk, Anibal Bregon, Jan Åslund, Mattias Krysander, Belarmino Pulido, and Gautam Biswas. Diagnosability analysis considering causal interpretations for differential constraints. *IEEE Transactions on Systems, Man, and Cybernetics – Part A: Systems and Humans*, 42(5):1216–1229, 2012.

[52] Erik Frisk, Dilek Düştegör, Mattias Krysander, and Vincent Cocquempot. Improving fault isolability properties by structural analysis of faulty behavior models: application to the DAMADICS benchmark problem. In *Proceedings of the 5th IFAC Symposium on Fault Detection, Supervision and Safety of Technical Processes (SAFEPROCESS'03)*, pages 1107–1112, Washington, D.C., USA, 2003.

[53] Erik Frisk and Mattias Krysander. Sensor placement for maximum fault isolability. In *Proceedings of the 18th International Workshop on Principles of Diagnosis (DX'07)*, pages 106–113, Nashville, USA, 2007.

[54] Erik Frisk, Mattias Krysander, Mattias Nyberg, and Jan Åslund. A toolbox for design of diagnosis systems. In *Proceedings of the 6th IFAC Symposium on Fault Detection, Supervision and Safety of Technical Processes (SAFEPROCESS'06)*, pages 657–662, Beijing, China, 2006.

[55] Esteban R. Gelso and Mogens Blanke. Structural analysis extended with active fault isolation - methods and algorithms. In *Proceedings of the 7th IFAC Symposium on Fault Detection, Supervision and Safety of Technical Processes (SAFEPROCESS'09)*, pages 597–602, Barcelone, Spain, 2009.

[56] Esteban R. Gelso, Sandra M. Castillo, and Joaquim Armengol. An algorithm based on structural analysis for model-based fault diagnosis. *Artificial Intelligence Research and Development. Frontiers in Artificial Intelligence and Applications*, 184:138–147, 2008.

[57] Toshiharu Hatanaka and Katsuji Uosaki. Optimal input design for discrimination of linear stochastic models based on Kullback-Leibler discrimination information measure. In *Proceedings of the IFAC Symposium on Identification and System Parameter Estimation (SYSID'88)*, pages 571–575, Beijing, China, 1988.

[58] Toshiharu Hatanaka and Katsuji Uosaki. Optimal auxiliary input for fault detection - frequency domain approach. In *Proceedings of the IFAC Symposium on Identification and System Parameter Estimation (SYSID'94)*, pages 1069–1074, Copenhagen, Denmark, 1995.

[59] Toshiharu Hatanaka and Katsuji Uosaki. Optimal auxiliary input for fault detection of systems with model uncertainty. In *Proceedings of the 1999 IEEE International Conference on Control Applications (CCA'99)*, pages 1436–1441, Kohala Coast, USA, 1999.

[60] Toshiharu Hatanaka and Katsuji Uosaki. Optimal auxiliary input for fault detection based on Kullback divergence. In *Proceedings of the 2000 IEEE International Conference on Industrial Electronics, Control and Instrumentation (IECON'00)*, pages 1731–1736, Nagoya, Japan, 2000.

[61] Rolf Isermann. *Fault-Diagnosis Systems*. Springer, 2006.

[62] Roozbeh Izadi-Zamanabadi. Structural analysis approach to fault diagnosis with application to fixed-wing aircraft motion. In *Proceedings of the 2002 American Control Conference (ACC'02)*, pages 3949–3954, Anchorage, USA, 2002.

[63] Roozbeh Izadi-Zamanabadi and Mogens Blanke. Structural analysis for diagnosis - the matching problem revisited. In *Proceedings of the 15h IFAC Word Congress*, pages 790–790, Barcelone, Spain, 2002.

[64] Roozbeh Izadi-Zamanabadi, Mogens Blanke, and Serajeddin Katebi. Cheap diagnosis using structural modelling and fuzzy-logic-based detection. *Control Engineering Practice*, 11(4):415–422, 2003.

[65] Roozbeh Izadi-Zamanabadi and Marcel Staroswiecki. A structural analysis method formulation for fault-tolerant control system design. In *Proceedings of the 39th IEEE Conference on Decision and Control (CDC'00)*, pages 4901–4902, Sydney, Australia, 2000.

[66] Festa Kerestecioglu and Martin B. Zarrop. *Issues of Fault Diagnosis for Dynamic Systems*, chapter 11. Input design for change detection, pages 315–338. Springer, London, 2000.

[67] Feza Kerestecioglu. *Change Detection and Input Design in Dynamical Systems*. Research Studies Press, Taunton, England, 1993.

[68] Feza Kerestecioglu and Ilker Cetin. Auxiliary input design for detecting changes towards partially known hypotheses. In *Proceedings of the 3rd IFAC Symposium on Fault Detection, Supervision and Safety for Technical Processes (SAFEPRO-CESS'97)*, pages 1027–1032, Kingston Upon Hull, United Kingdom, 1997.

[69] Feza Kerestecioglu and Ilker Cetin. Auxiliary signal design for detecting changes towards unknown hypotheses. In *Proceedings of the 1997 IEEE International Symposium on Intelligent Control*, pages 297–302, Istanbul, Turkey, 1997.

[70] Feza Kerestecioglu and Ilker Cetin. Optimal input design for the detection of changes towards unknown hypotheses. *International Journal of Systems Science*, 35(7):435–444, 2004.

[71] Feza Kerestecioglu and Martin B. Zarrop. Optimal input design for change detection in dynamical systems. In *Proceedings of the 1st European Control Conference (ECC'91)*, pages 321–326, Grenoble, France, 1991.

[72] Feza Kerestecioglu and Martin B. Zarrop. Input design for detection of abrupt changes in dynamical systems. *International Journal of Control*, 59(4):1063–1084, 1994.

[73] Markus Koslowski. MATLAB toolbox for structural analysis of dynamical systems. Internship report, Robert Bosch GmbH, 2010.

[74] Mattias Krysander. *Design and Analysis of Diagnosis Systems Using Structural Methods*. PhD thesis, Linköpings universitet, 2006.

[75] Mattias Krysander, Jan Åslund, and Mattias Nyberg. An efficient algorithm for finding over-constrained sub-systems for construction of diagnostic tests. In *Proceedings of the 16th International Workshop on Principles of Diagnosis (DX'05)*, Pacific Grove, USA, 2005.

[76] Mattias Krysander, Jan Åslund, and Mattias Nyberg. An efficient algorithm for finding minimal overconstrained subsystems for model-based diagnosis. *IEEE Transactions on Systems Man and Cybernetics - Part A: Systems and Humans*, 38(6):197–206, 2008.

[77] Mattias Krysander and Erik Frisk. Sensor placement for fault diagnosis. *IEEE Transactions on Systems, Man, and Cybernetics, Part A: Systems and Humans*, 38(6)(1):1398 –1410, 2008.

[78] Mattias Krysander and Nyberg. Structural analysis utilizing MSS sets with application to a paper plant. In *Proceedings of the 13th International Workshop on Principles of Diagnosis (DX'02)*, Semmering, Austria, 2002.

[79] Mattias Krysander and Mattias Nyberg. Structural analysis for fault diagnosis of DAE systems utilizing MSS sets. In *Proceedings of the 15th IFAC World Congress*, pages 753–753, Barcelone, Spain, 2002.

[80] Mattias Krysander, Jan Åslund, and Erik Frisk. A structural algorithm for finding testable sub-models and multiple fault isolability analysis. In *Proceedings of the 21st International Workshop on the Principles of Diagnosis (DX'10)*, Portland, USA, 2010.

[81] Morten Laurensen, Mogens Blanke, and Dilek Düştegör. Fault diagnosis of a water for injection system using enhanced structural isolation. *International Journal of Applied Mathematics and Computer Science*, 18(4):593–603, 2008.

[82] Jan Lunze, Thomas Steffen, and Uta Riedel. Fault diagnosis of dynamical systems using state-set observers. In *Proceedings of the 14th International Workshop on Principles of Diagnosis (DX'03)*, pages 1131–1136, Washington, D.C., USA, 2003.

[83] Fatiha Nejjari, Ramon Sarrate, and Albert Rosich. Optimal sensor placement for fuel cell system diagnosis using bilp formulation. In *Proceedings of the 18th Mediterranean Conference on Control and Automation (MED'10)*, pages 1296–1301, Marrakech, Morocco, 2010.

[84] Henrik Niemann. Fault tolerant control based on active fault diagnosis. In *Proceedings of the 2005 American Control Conference (ACC'05)*, pages 2224–2229, Portland, USA, 2005.

[85] Henrik Niemann. Active fault diagnosis in closed-loop uncertain systems. In *Proceedings of the 6th IFAC Symposium on Fault Detection Supervision and Safety for Technical Processes (SAFEPROCESS'06)*, pages 631–636, Beijing, China, 2006.

[86] Henrik Niemann. A setup for active fault diagnosis. *IEEE Transactions on Automatic Control*, 51(9):1572–1578, 2006.

[87] Henrik Niemann and Niels K. Poulsen. Active fault diagnosis in closed-loop systems. In *Proceedings of the 16th IFAC World Congress*, pages 1876–1876, Prague, Czech Republic, 2005.

[88] Henrik Niemann and Niels K. Poulsen. Active fault diagnosis for systems with reduced model information. In *Proceedings of the 7th IFAC Symposium on Fault Detection, Supervision and Safety of Technical Processes (SAFEPROCESS'09)*, pages 965–970, Barcelone, Spain, 2009.

[89] Henrik Niemann, Niels K. Poulsen, and Mikkel Ask Buu Bækgaard. A multi-model approach for system diagnosis. In *Proceedings of the 2007 American Control Conference (ACC'07)*, pages 2539–2544, New York City, USA, 2007.

[90] Henrik Niemann, Niels K. Poulsen, Henrik Parbo, and Michael L. Nielsen. Active system monitoring applied on wind turbines. In *Proceedings of the Nordic Wind Power Conference (NWPC'09)*, Bornholm, Denmark, 2009.

[91] Ramine Nikoukhah. Guaranteed active failure detection and isolation for linear dynamical systems. *Automatica*, 34:1345–1358, 1998.

[92] Ramine Nikoukhah and Stephen L. Campbell. Active failure detection: Auxiliary signal design and on-line detection. In *Proceedings of the 10th Mediterranean Conference on Control and Automation (MED'02)*, Lisbon, Portugal, 2002.

[93] Ramine Nikoukhah and Stephen L. Campbell. Auxiliary signal design for active failure detection in uncertain linear systems with a priori information. *Automatica*, 42:219–228, 2006.

[94] Ramine Nikoukhah and Steven L. Campbell. The design of auxiliary signals for robust active failure detection in uncertain systems. In *Proceedings of the 15th International Symposium on Mathematical Theory of Networks and Systems (MTNS'02)*, South Bend, USA, 2002.

[95] Ramine Nikoukhah and Steven L. Campbell. Auxiliary signal design for robust active failure detection: the general cost case. In *Proceedings of the 5th IFAC Symposium on Fault Detection, Supervision and Safety of Technical Processes (SAFEPROCESS'03)*, Washington, D.C., USA, 2003.

[96] Ramine Nikoukhah and Steven L. Campbell. Auxiliary signal design in uncertain systems with known inputs. In *Proceedings of the 16th IFAC World Congress*, Prague, Czech Republic, 2005.

[97] Ramine Nikoukhah and Steven L. Campbell. Robust detection of incipient faults: an active approach. In *Proceedings of the 14th Mediterranean Conference on Control and Automation (MED'06)*, Ancona, Italy, 2006.

[98] Ramine Nikoukhah, Steven L. Campbell, and Francois Delebecque. Auxiliary signal design for failure detection in uncertain systems. In *Proceedings of the 9th Mediterranean Conference on Control and Automation (MED'01)*, Dubrovnik, Croatia, 2001.

[99] Ramine Nikoukhah, Steven L. Campbell, and Kimberly J. Drake. An active approach for detection of incipient faults. *International Journal of Systems Science*, 41(2):241–257, 2010.

[100] Niels K. Poulsen and Henrik Niemann. Stochastic change detection based on an active fault diagnosis approach. In *Proceedings of the 46th IEEE Conference on Decision and Control (CDC'07)*, pages 346–351, New Orleans, USA, 2007.

[101] Niels K. Poulsen and Henrik Niemann. Active fault diagnosis based on stochastic tests. *International Journal of Applied Mathematics and Computer Science*, 18(4):487–496, 2008.

[102] Belarmino Pulido and Carlos Alonso González. Possible conflicts, ARRs, and conflicts. In *Proceedings of the 13th International Workshop on the Principles of Diagnosis (DX'02)*, pages 122–127, Semmering, Austria, 2002.

[103] Belarmino Pulido and Carlos Alonso González. Possible conflicts: A compilation technique for consistency-based diagnosis. *IEEE Transactions on Systems, Man, and Cybernetics - Part B: Cybernetics*, 34(5):2192–2206, 2004.

[104] Harald Renninger and Martin Konieczny. Dynamische Priorisierung von Prüfschritten in der Werkstattdiagnose. *at - Automatisierungstechnik*, 55:28–34, 2007.

[105] Robert N. Riggins and William B. Ribbens. Designed inputs for detection and isolation of failures in the state transition matrices of dynamic systems. *IEEE Transactions on Control Systems Technology*, 5(2):149–162, 1997.

[106] Albert Rosich. Sensor placement for fault detection and isolation based on structural models. In *Proceedings of the 8th IFAC Symposium on Fault Detection, Supervision and Safety of Technical Processes (SAFEPROCESS'12)*, pages 391–396, Mexico City, Mexico, 2012.

[107] Albert Rosich, Erik Frisk, Jan Åslund, Ramon Sarrate, and Fatiha Nejjari. Sensor placement for fault diagnosis based on causal computations. In *Proceedings of the 7th IFAC Symposium on Fault Detection, Supervision and Safety of Technical Processes (SAFEPROCESS'09)*, pages 402–407, Barcelone, Spain, 2009.

[108] Albert Rosich, Erik Frisk, Jan Åslund, Ramon Sarrate, and Fatiha Nejjari. Fault diagnosis based on causal computations. *IEEE Transactions on Systems, Man and Cybernetics, Part A: Systems and Humans*, 42(2):371–381, 2012.

[109] Albert Rosich, Ramon Sarrate, and Fatiha Nejjari. Optimal sensor placement for FDI using binary integer linear programming. In *Proceedings of the 20th International Workshop on Principles of Diagnosis (DX'09)*, pages 235–241, Stockholm, Sweden, 2009.

[110] Albert Rosich, Ramon Sarrate, Vicenc Puig, and Teresa Escobet. Efficient optimal sensor placement for model-based FDI using an incremental algorithm. In *Proceedings of the 46th IEEE Conference on Decision and Control, (CDC'07)*, pages 2590–2595, New Orleans, USA, 2007.

[111] Albert Rosich, Abed Alrahim Yassine, and Stéphane Ploix. Efficient optimal sensor placement for structural model based diagnosis. In *Proceedings of the 21st International Workshop on the Principles of Diagnosis (DX'10)*, Portland, USA, 2010.

[112] Ramon Sarrate, Fatiha Nejjari, and Albert Rosich. Model-based optimal sensor placement approaches to fuel cell stack system fault diagnosis. In *Proceedings of the 8th IFAC Symposium on Fault Detection, Supervision and Safety of Technical Processes (SAFEPROCESS'12)*, pages 96–101, Mexico City, Mexico, 2012.

[113] Ramon Sarrate, Vicenc Puig, Teresa Escobet, and Albert Rosich. Optimal sensor placement for model-based fault detection and isolation. In *Proceedings of the 46th IEEE Conference on Decision and Control (CDC'07)*, pages 2584–2589, New Orleans, USA, 2007.

[114] Jakob Stoustrup and Henrik Niemann. Active fault diagnosis by temporary destabilization. In *Proceedings of the 6th IFAC Symposium on Fault Detection, Supervision and Safety for Technical Processes (SAFEPROCESS'06)*, pages 563–568, Beijing, China, 2006.

[115] Jakob Stoustrup and Henrik Niemann. Active fault diagnosis by controller modification. *International Journal of Systems Science*, 41:925–936, 2010.

[116] Carl Svärd. *Methods for Automated Design of Fault Detection and Isolation Systems with Automotive Applications*. PhD thesis, Linköpings universitet, 2012.

[117] Carl Svärd and Mattias Nyberg. Residual generators for fault diagnosis using computation sequences with mixed causality applied to automotive systems. *IEEE Transactions on Systems, Man, and Cybernetics – Part A: Systems and Humans*, 40(6):1310–1328, 2010.

[118] Carl Svärd, Mattias Nyberg, and Erik Frisk. A greedy approach for selection of residual generators. In *Proceedings of the 22nd International Workshop on Principles of Diagnosis (DX'11)*, Murnau, Germany, 2011.

[119] Kelly Sweetingham. *Auxiliary Signal Design for Fault Detection in Nonlinear Systems*. PhD thesis, North Carolina State University, 2008.

[120] Louise Travé-Massuyès. Bridging technologies for diagnosis. In *Proceedings of the 8th IFAC Symposium on Fault Detection, Supervision and Safety of Technical Processes (SAFEPROCESS'12)*, pages 361–372, Mexico City, Mexico, 2012.

[121] Louise Travé-Massuyès, Teresa Escobet, and Xavier Olive. Diagnosability analysis based on component supported analytical redundancy relations. *IEEE Transactions on Systems, Man and Cybernetics, Part A: Systems and Humans*, 36(6):1146–1160, 2006.

[122] Louise Travé-Massuyès, Teresa Escobet, and Stefan Spanache. Diagnosability analysis based on component supported analytical redundancy relations. In *Proceedings of the 16th IFAC World Congress 2005*, pages 897–902, Prague, Czech Republic, 2005.

[123] Katsuji Uosaki and Toshiharu Hatanaka. Optimal input design for autoregressive model discrimination with constrained output variance. *IEEE Transactions on Automatic Control*, 29(4):348–350, 1984.

[124] Katsuji Uosaki and Toshiharu Hatanaka. Optimal auxiliary input for fault detection and fault diagnosis. In *Proceedings of the 1996 IEEE International Symposium on Computer-Aided Control System Design (CACSD'96)*, Dearborn, USA, 1996.

[125] Katsuji Uosaki and Toshiharu Hatanaka. Optimal auxiliary input design for fault detection based on Kullback divergence. In *Proceedings of the 5th Asian Control Conference*, pages 428–434, Melbourne, Australia, 2004.

[126] Katsuji Uosaki, N. Takata, and Toshiharu Hatanaka. Optimal auxiliary input for on-line fault detection and fault diagnosis. In *Proceedings of the Automatic Control World Congress 1993*, pages 441–446, Sydney, Australia, 1993.

[127] Wang Xiazhong and Roger M. Cooke. Optimal inspection sequence in fault diagnosis. *Reliability Engineering and System Safety*, 37(3):207–210, 1992.

[128] Wang Xiazhong and Rune Reinertsen. General inspection strategy for fault diagnosis-minimizing the inspection costs. *Reliability Engineering and System Safety*, 48:191–197, 1995.

[129] Abed Alrahim Yassine, Albert Rosich, and Stéphane Ploix. An optimal sensor placement algorithm taking into account diagnosability specifications. In *Proceedings of the 17th IEEE International Conference on Automation, Quality and Testing, Robotics (AQTR)*, Cluj-Napoca, Romania, 2010.

[130] Xue J. Zhang. *Auxiliary Signal Design in Fault Detection and Diagnosis*. Springer, 1989.

Curriculum vitae

Name	Michael Volker Ungermann
Date of birth	May 29th, 1982
Place of birth	Hanau, Germany

Education

08/2008 - present   Ruhr-Universität Bochum, Germany
Institute for Automation and Computer Control (ATP)
PhD-Student in Control Theory

10/2002 - 05/2008   Technische Universität Darmstadt, Germany
Studies in Electrical Engineering
German-French Double-Degree (Dipl.-Ing. 2008)

09/2005 - 02/2007   Institut national polytechnique de Grenoble, France
Studies in Automation & Image and Signal Processing
German-French Double-Degree (Ingénieur diplômé 2008)

07/2001 - 03/2002   Bundeswehr, Lebach, Germany
Military Service

08/1992 - 06/2001   Grimmelshausen Gymnasium Gelnhausen, Germany
Secondary school

08/1988 - 06/1992   Grundschule Eidengesäß, Germany
Primary school

Profession and Internships

09/2011 - present   Robert Bosch GmbH, Schwieberdingen, Germany
R&D-Engineer: function development for electrical vehicles

08/2008 - 08/2011   Robert Bosch GmbH, Schwieberdingen, Germany
PhD-program: automatic tests for service diagnosis

05/2007 - 10/2007   Robert Bosch India ltd., Bangalore, India
Internship: feasibility study for automotive SW development

04/2002 - 07/2002   Indramat GmbH, Lohr am Main, Germany
Basic internship